AUTOMATIC GAIN CONTROL: A PRACTICAL APPROACH TO ITS ANALYSIS AND DESIGN

With Applications to Radar

by Richard Smith Hughes

Second Edition

with corrections by Greg Easter

WEXFORD PRESS

2007

Preface

Automatic gain control is used in virtually every radio and TV receiver and in various radio frequency test instruments. In addition, many radar receivers employ AGC prior to signal processing. One could assume, with the wide applications for AGC, that the design principles would be widely publicized. This, however, is not the case. There appears to be no single reference devoted to this important topic.

This book presents all pertinent aspects of AGC analysis and design as applied to radar-type receivers, from static regulation to dynamic regulation (modulation reduction), to loop bandwidth and rise time. The theory presented is equally valid for communication and TV receivers. The effect of detector type, linear and square law, on AGC action is covered in detail, and the theory presented is verified with a simple test circuit. The results are summarized in easy to use design charts (Figures 20 and 21).

Several practical applications are presented in Chapter 2, using the theory developed in Chapter 1. These applications include scanning beam radar AGC, conical scan radar AGC, and a simple AGC loop employing only one operational amplifier. A method to minimize the effect of input signal step on AGC loop rise time is also presented. These examples are not meant to be an exhaustive design dissertation, but are presented only as possible solutions. The individual designer still has complete control in the circuit design, but now at last through the techniques presented here, he has an analytical head start in optimizing the AGC loop for his individual needs.

Richard Smith Hughes
China Lake, Calif.

Acknowledgment

I am indebted to many people in the preparation of this book: Dr. H. Wade Swinford for his many stimulating discussions and suggestions; Paul J. Hilliard, who constructed and tested all of the circuits presented; D. J. Russell, Head of the Electronic Warfare Department, who created the necessary practical and academic atmosphere that made this work possible.

Contents

1

Basics of Automatic Gain Control

INTRODUCTION

Many radars employ automatic gain control (AGC) to normalize the received signal prior to signal processing (i.e., range tracking, signal acquisition, direction finding, etc.) as shown in Figure 1. As the received input signal varies, the input to the intermediate frequency (IF) amplifier changes by the same amount. The AGC loop notes the changes and varies the gain of the IF amplifier in such a way that the output of the detector remains constant (the AGC voltage could also vary the gain of the radio frequency (RF) amplifier).

Figure 2 illustrates the basic components of AGC tracking loops (continuous wave (CW) inputs will be assumed for now; however, operation with pulse inputs is basically the same and will be covered later). The only difference between Figures 2a and 2b is that one uses a low-pass filter (LPF) and the other an integrator. The differences between these two techniques will be discussed shortly.

The configurations of Figure 2 are quite general in that most AGC loops can be reduced to the components illustrated (necessary pulse stretching and timing circuitry for pulse AGC operation are not shown in the interest of simplicity). The input signal, P_{in}(dBm),* is amplified by the variable gain IF amplifier. The variable gain IF output may be amplified or attenuated, depending on the wanted normalized output

* For typographical reasons expressions such as $P_{\text{in}}\big|^{\text{dBm}}$ and $P_{IF}\big|_{SL}^{\text{dBm}}$ are presented as P_{in}(dBm) and $P_{IF,SL}$(dBm) in this book.

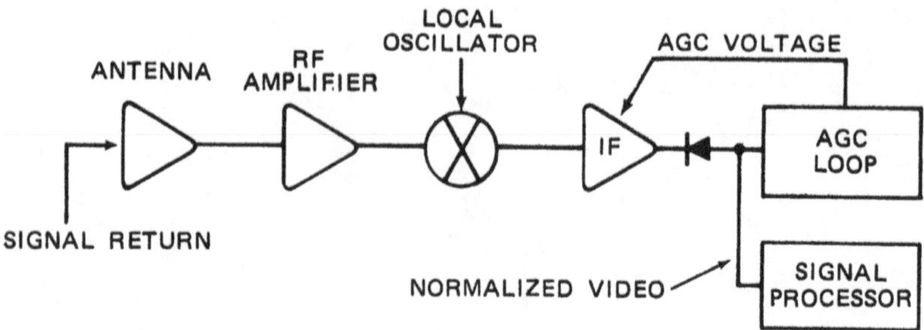

FIGURE 1. Basic Radar Receiver.

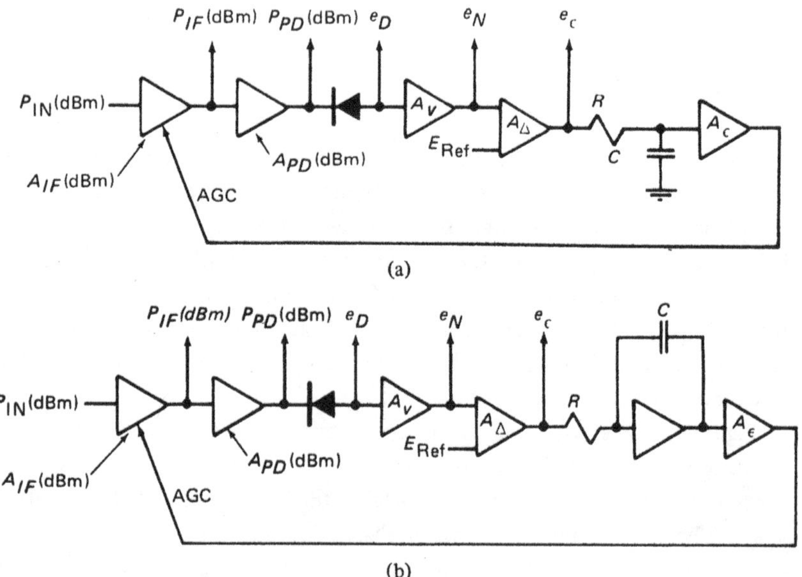

FIGURE 2. Basic Components of Automatic Gain Control (AGC) Tracking Loops. (a) Low-pass filter, (b) integrator.

power, $P_{o,N}$(dBm), and detector input power, $P_{PD,N}$(dBm). The detected signal is amplified by the video amplifier, A_v, compared with a reference voltage, E_{Ref}, and again amplified by the error amplifier, A_ϵ. The resultant voltage drives the variable gain IF AGC input.

If P_{in}(dBm) should increase, P_{o}(dBm), P_{PD}(dBm), and thus the normalized video voltage, e_N, would increase, increasing the AGC voltage, and thus decreasing the gain until $e_N = E_{\text{Ref}}$.

Three basic parameters define the operation on an AGC loop:

1. Static regulation is the capability to compress large input variations into small output variations. This is the same concept as regulation in a regulated power supply. This compressed output variation, ΔP_o(dBm), divided by the input variation, ΔP_{in}(dBm) is called the compression ratio (*CR*).

$$CR = \frac{\Delta P_{IF}(\text{dBm})}{\Delta P_{in}(\text{dBm})} \tag{1-1}$$

2. Dynamic regulation is the capability of an AGC loop to reduce the dynamic input modulation appearing at the output, P_{IF}(dBm), and is called the input modulation reduction (IMR). This quality is especially important in conical scan radars. IMR is dependent on the loop gain (LG) of the AGC loop and may be given as

$$IMR = \frac{1}{1+LG} \tag{1-2}$$

The AGC loop output modulation (M_o) may be given as

$$M_o = (IMR)M_I \tag{1-3}$$

where M_I is the input modulation. Thus a small IMR is wanted (large loop gain).

3. Loop rise time (τ_r) is the 10 to 90% loop-response time due to a step change in input power.

These three parameters will be discussed thoroughly in the next sections.

Static Regulation

Assume that the variable gain IF amplifier has a variable gain characteristic as illustrated in Figure 3. (Most variable gain IF amplifiers

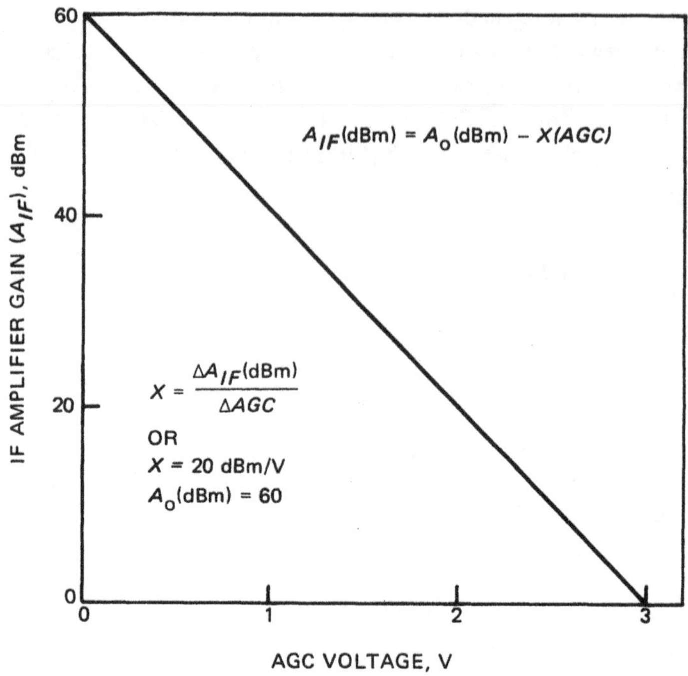

FIGURE 3. IF Amplifier Gain Versus AGC Voltage.

have a fairly linear relationship between gain (dBm) and AGC voltage.) The equation relating IF gain to AGC voltage is

$$A_{IF}(\text{dBm}) = A_o(\text{dBm}) - X(AGC) \tag{1-4}$$

where

X = variable gain slope in dBm/V
A_o(dBm) = maximum gain

Defining gain in decibels with reference to 1 milliwatt (dBm) bears some explanation. AGC loops involve power levels and ratios in the IF and detector portion and voltage levels and ratios in the video (or postdetection) portion. To avoid confusion, power gains and amplitudes will be specified in dBm and voltage gains and amplitudes in decibels with reference to 1 volt (dBV). Appendix A summarizes power-voltage relationships.

One primary function for an AGC loop is that it keep the output power, P_o(dBm) (or video voltage, e_N) normalized to within a specified amount, ΔP_{IF}(dBm), or Δe_N(dBV), despite large variations in the input power, ΔP_{in}(dBm). The predetector amplifier may be necessary to ensure that the detector is operated at the desired level, linear or square law (see Appendix B for detector characteristics pertinent to AGC design).

The static regulation characteristics of the LPF AGC (Figure 2a) will now be discussed. Compression ratio has been defined in Equation 1-1. In Appendix C* we solve for ΔP_{IF}(dBm) as a function of detector type, linear (*Lin*) or square law (*SL*), and the results are

$$\Delta P_{IF,SL}(\text{dBm}) = 10 \log \left[\frac{\Delta P_{in}(\text{dBm})}{X A_\Delta A_\epsilon e_N} + 1 \right] \tag{1-5}$$

$$\Delta P_{IF,Lin}(\text{dBm}) = 20 \log \left[\frac{\Delta P_{in}(\text{dBm})}{X A_\Delta A_\epsilon e_N} + 1 \right] \tag{1-6}$$

Thus, to minimize the change in output power, e_N, A_Δ and A_ϵ must be made as large as practical (X is assumed constant for a given IF amplifier).

Solving Equations 1-5 and 1-6 for $A_\Delta A_\epsilon e_N$,

$$(A_\Delta A_\epsilon e_N)_{SL} = \frac{\Delta P_{in}(\text{dBm})}{X\left(10^{\frac{\Delta P_{IF}(\text{dBm})}{10}} - 1\right)} \tag{1-7}$$

$$(A_\Delta A_\epsilon e_N)_{Lin} = \frac{\Delta P_{in}(\text{dBm})}{X\left(10^{\frac{\Delta P_{IF}(\text{dBm})}{20}} - 1\right)} \tag{1-8}$$

* Derivations are given in Appendix C.

Thus, since we know X and ΔP_{IF}(dBm), the necessary $A_\Delta A_\epsilon e_N$ may be found. The usefulness of these equations is illustrated by the following example.

Example: Assume that a maximum deviation in output power of 1 dBm for an input power variation of 50 dBm is wanted. The variable gain slope, X, is 20 dBm/V.

a. Square law detector (Equation 1-7):

$$CR = 1/50 = 0.02 \tag{1-9}$$

$$(A_\Delta A_\epsilon e_N)_{SL} = \frac{50}{20\left(10^{\frac{1}{10}} - 1\right)} \tag{1-10}$$

or

$$(A_\Delta A_\epsilon e_N)_{SL} = 9.66 \tag{1-11}$$

since

$$e_N = A_\nu e_{D,N}$$

$$(A_\Delta A_\epsilon A_\nu e_{D,N})_{SL} = 9.66 \tag{1-12}$$

Assuming that $e_{D,N}$ = 10 mV (a reasonable value for a square law detector),

$$A_\Delta A_\epsilon A_\nu = 965 \tag{1-13}$$

or, in terms of dBV,

$$(A_\Delta A_\epsilon A_\nu)_{SL}(\text{dBV}) = 20 \log 965 = 59.7 \text{ dBV} \tag{1-14}$$

b. Linear detector (Equation 1-8)

$$(A_\Delta A_\epsilon e_N)_{Lin} = \frac{50}{20\left(10^{\frac{1}{20}} - 1\right)} \tag{1-15}$$

or

$$(A_\Delta A_\epsilon e_N)_{Lin} = 20.49 \tag{1-16}$$

Assuming that $e_{D,N}$ = 150 mV (again a reasonable value for a linear detector),

$$A_\Delta A_\epsilon A_\nu = 136.6 \tag{1-17}$$

or

$$(A_\Delta A_\epsilon A_\nu)_{Lin}(\text{dBV}) = 20 \log 136.6 = 42.7 \text{ dBV} \tag{1-18}$$

It is apparent that a linear detector requires less gain than a square law detector. This may be useful if the AGC loop must be DC-coupled (as would be the case for a continuous wave AGC loop); however, the IF amplifier (or predetector amplifier) must be capable of providing the necessary power to keep the detector in the linear region.

To illustrate the discussion presented thus far, consider the following:

Minimum input for AGC action (or AGC delay)	$P_{\text{in,min}}(\text{dBm}) = -70$ dBm
Minimum output under AGC action	$P_{IF,\text{min}}(\text{dBm}) = -0.5$ dBm
Maximum output under AGC action	$P_{IF,\text{max}}(\text{dBm}) = +0.5$ dBm
Maximum input for AGC action (or AGC dropout level)	$P_{\text{in,max}}(\text{dBm}) = -20$ dBm

The input dynamic range is

$$\Delta P_{\text{in}}(\text{dBm}) = P_{\text{in,max}}(\text{dBm}) - P_{\text{in,min}}(\text{dBm}) = 50 \text{ dBm} \tag{1-19}$$

and the output dynamic range is

$$\Delta P_{\text{o}}(\text{dBm}) = P_{IF,\text{max}}(\text{dBm}) - P_{IF,\text{min}}(\text{dBm}) = 1 \text{ dBm} \tag{1-20}$$

Thus the compression ratio is

$$CR = \frac{\Delta P_{IF}(\text{dBm})}{\Delta P_{\text{in}}(\text{dBm})} = 0.02 \tag{1-21}$$

or, the output increases 0.02 dBm for each 1 dBm increase on the input.

Figure 4 illustrates the characteristics of the AGC system just presented.

This discussion has been for the AGC of Figure 2a. Figure 2b illustrates an AGC loop incorporating a true integrator. A true integrator has near infinite gain at low frequencies; therefore, under normalized, nonmodulation inputs, $e_\epsilon = 0$. Thus,

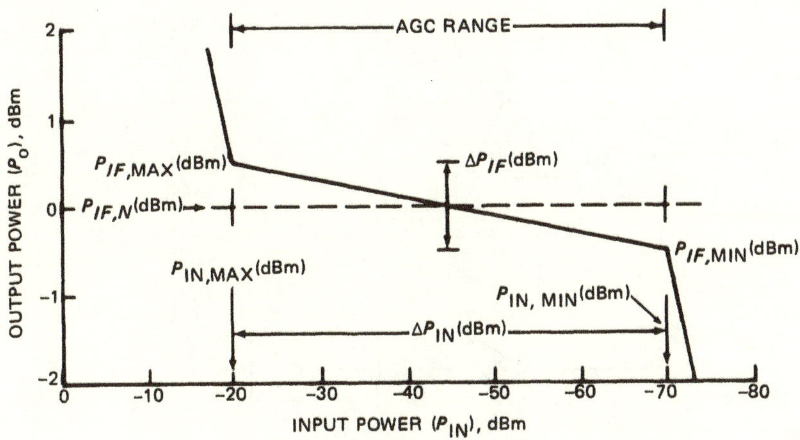

FIGURE 4. Typical AGC Characteristics.

$$\Delta P_{IF}(\text{dBm}) \cong 0 \tag{1-22}$$

and perfect regulation is obtained (further in the chapter the theoretical equations presented will be verified with a practical example).

The next section will present the behavior of the AGC loop illustrated in Figure 2, under input modulation conditions.

Dynamic Regulation

Figure 5 illustrates the AGC loop of Figure 2a in classical feedback form. Using conventional feedback theory, Oliver[1] has shown that

$$\frac{\Delta e_{IF}(PP)}{e_{IF}(PP)} = \frac{1}{1+AB}\left(\frac{\Delta e_{\text{in}}(PP)}{e_{\text{in}}(PP)}\right) \tag{1-23}$$

[1] B. M. Oliver. "Automatic Volume Control as a Feedback Problem," *Proceedings of the IRE.* April 1948, pp. 466-73.

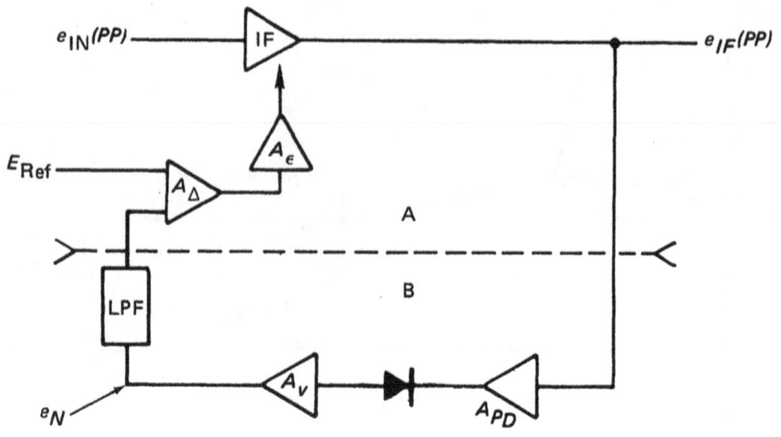

FIGURE 5. Block Diagram of Feedback Amplifier.

where

$e_{IF}(PP)$ = peak-to-peak IF output voltage

$\frac{\Delta e_{IF}(PP)}{e_{IF}(PP)}$ (100) = percent of output modulation (M_o)

$\frac{\Delta e_{in}(PP)}{e_{in}(PP)}$ (100) = percent of input modulation (M_I)

AB = loop gain (LG)

Now Equation 1-23 may be written as (assuming $LG >> 1$)

$$M_{IF} = \frac{M_I}{LG} \tag{1-24}$$

or the output modulation, of the IF amplifier, is reduced by the inverse of the loop gain. Thus the input modulation reduction, IMR, is (for large loop gains)

$$IMR \cong \frac{1}{LG} \tag{1-25}$$

and

$$M_{IF} = IMR\,(M_I) \tag{1-26}$$

The loop gain may be found (for constant input power) by breaking the AGC input to the variable gain IF amplifier and modulating this voltage. The loop gain is thus the modulated output voltage, $\Delta AGC'$, divided by the modulated AGC voltage, ΔAGC, as illustrated in Figure 6.*

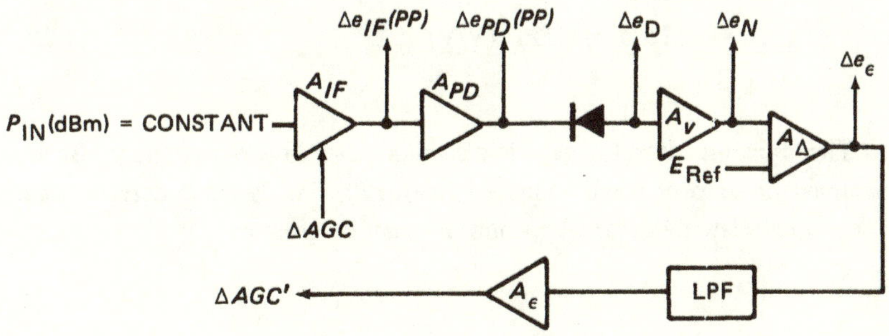

FIGURE 6. Method of Finding Loop Gain.

$$LG = \frac{\Delta AGC'}{\Delta AGC} \tag{1-27}$$

or

$$LG = A_{AGC} A_{PD} A_D A_v A_\Delta A_\epsilon \tag{1-28}$$

where

* The effect of the low-pass filter (LPF) will be neglected Ideally this sets the frequency response of the loop.

A_{AGC} = dynamic AGC gain, $\Delta e_{IF}(PP)/\Delta AGC$ (V/V)

A_{PD} = predetector gain, $\Delta e_{PD}(PD)/\Delta e_{IF}(PP)$ (V/V)

A_D = dynamic detector gain, $\Delta e_D/\Delta e_{PD}(PP)$ (V/V)

A_v = video gain, $\Delta e_N/\Delta e_D$ (V/V)

A_Δ = differencing gain $\Delta e_\epsilon/\Delta e$ (V/V)

A_ϵ = error gain $\Delta AGC'/\Delta e_\epsilon$ (V/V)

The dynamic AGC gain is a nonlinear quantity, but for small values of ΔAGC may be given as (Appendix D)

$$A_{AGC} = 0.115 \; X \; e_{IF}(PP) \quad \text{(V/V)} \tag{1-29}$$

The dynamic detector gain is also nonlinear; however, it may also be assumed linear over small values of $\Delta e_{PD}(PP)$. The dynamic detector gain for a square law detector (Appendix B) may be given as

$$A_{PD} = 2K_{SL} e_{PD}(PP) \quad \text{(V/V)} \tag{1-30}$$

where K_{SL} is a detector constant.

Substituting Equations 1-29 and 1-30 into 1-28,

$$LG_{SL} = [0.115 \; X \; e_{IF}(PP)]A_{PD} \; 2K_{SL} e_{PD}(PP) \; A_v A_\Delta A_\epsilon \tag{1-31}$$

However, since

$$e_{IF}(PP) = e_{PD}(PP)/A_{PD} \tag{1-32}$$

Equation 1-31 may now be written as

$$LG_{SL} = 0.115\, X[e_{PD}(PP)]^2 2K_{SL}A_v A_\Delta A_\epsilon \tag{1-33}$$

The detector output, e_D, is (Appendix B)

$$e_D = K_{SL}[e_{PD}(PP)]^2$$

thus, under normalized conditions, $e_D = e_{D,N}$, and

$$LG_{SL} = 0.23\, XA_v A_\Delta A_\epsilon e_{D,N} \tag{1-34}$$

or, since

$$e_N = e_{D,N}A_v \tag{1-35}$$

$$LG_{SL} = 0.23XA_\Delta A_\epsilon e_N \tag{1-36}$$

Equation 1-36 is simple, but very accurate, as will be shown. Using the same methods, but for a linear detector (Appendix B),

$$A_{D,Lin} = K_{Lin} \tag{1-37}$$

and

$$e_{D,Lin} = K_{Lin}\, e_{PD}(PP) \tag{1-38}$$

Thus the loop gain becomes

$$LG_{Lin} = 0.12\ XA_{\Delta}A_{\epsilon}e_N \tag{1-39}$$

The loop gain will now be found for the problem described in the example on pages 8 and 9.

a. Square law detector: $A_{\Delta}A_{\epsilon}e_N = 9.66$
   $X = 20$ dBm/V

or

$$LG_{SL} = (0.23)(20)\ 9.66 \tag{1-40}$$

$$LG_{SL} = 44.4 \quad (32.9 \text{ dBV}) \tag{1-41}$$

b. Linear detector: $A_{\Delta}A_{\epsilon}e_N = 20.49$
   $X = 20$ dBm/V

$$LG_{Lin} = (0.12)\ (20)\ (20.49) \tag{1-42}$$

or

$$LG_{Lin} = 49.2 \quad (33.8 \text{ dBV}) \tag{1-43}$$

From Equations 1-5 and 1-36 it can be seen that the loop gain and static regulation (ΔP_{IF}(dBm)) are dependent on $XA_{\Delta}A_{\epsilon}e_N$; thus, for a given IF amplifier $A_{\Delta}A_{\epsilon}$ can be maximized to give the necessary loop gain:

$$\Delta P_{IF,SL}(\text{dBm}) = 10 \log \left[\frac{\Delta P_{\text{in}}(\text{dBm})}{XA_{\Delta}A_{\epsilon}e_N} + 1 \right] \tag{1-44}$$

and

$$LG_{SL} = 0.23\ XA_{\Delta}A_{\epsilon}e_N \tag{1-45}$$

Solving Equation 1-45 for $A_{\Delta}A_{\epsilon}e_N$,

$$A_{\Delta}A_{\epsilon}e_N = \frac{LG_{SL}}{0.23X} \tag{1-46}$$

and substituting into Equation 1-44,

$$\Delta P_{IF,SL}(\text{dBm}) = 10 \log \left[\frac{0.23\ \Delta P_{\text{in}}(\text{dBm})}{LG} + 1 \right] \tag{1-47}$$

For the linear detector this equation becomes

$$\Delta P_{IF,Lin}(\text{dBm}) = 10 \log \left[\frac{0.12\ \Delta P_{\text{in}}(\text{dBm})}{LG} + 1 \right] \tag{1-48}$$

Thus the loop gain uniquely determines the static regulation. Conversely, the static regulation uniquely determines the loop gain, as shown below.

Solving Equation 1-44 for $A_{\Delta}A_{\epsilon}e_N$,

$$A_{\Delta}A_{\epsilon}e_{N,SL} = \frac{\Delta P_{\text{in}}(\text{dBm})}{X\left(10^{\frac{\Delta P_{IF}(\text{dBm})}{10}} - 1\right)} \tag{1-49}$$

Substituting Equation 1-49 into 1-45,

$$LG_{SL} = 0.23 \left[\frac{\Delta P_{\text{in}}(\text{dBm})}{10^{\frac{\Delta P_{IF}(\text{dBm})}{10}} - 1} \right] \tag{1-50}$$

and for the linear detector,

$$LG_{Lin} = 0.12 \left[\frac{\Delta P_{\text{in}}(\text{dBm})}{10^{\frac{\Delta P_{IF}(\text{dBm})}{20}} - 1} \right] \tag{1-51}$$

Thus if one is designing for a given static regulation, he has no control over loop gain, and vice versa.

The dynamic regulation for the true integrator AGC loop of Figure 2b is similar to the low-pass-filter loop, except that the loop gain must be multiplied by the frequency-dependent gain of the integrator, A_{Int},

$$A_{Int} = Z_F/R \tag{1-52}$$

where

$$Z_F = -j/2\pi fC \tag{1-53}$$

or

$$A_{Int} = 1/2\pi fCR \ \angle{-90} \text{ degrees} \tag{1-54}$$

where the −90-degree phase shift is due to the $-j$ term of the capacitor. Thus,

$$LG_{SL} = \frac{0.036\ XA_{\Delta}A_{e}e_{N}}{fCR} \tag{1-55}$$

and

$$LG_{Lin} = \frac{0.018\ XA_{\Delta}A_{\epsilon}e_N}{fCR} \tag{1-56}$$

The frequency response for the low-pass-filter loop will be determined by the low-pass filter (assuming that the open loop frequency, neglecting the filter, is much longer than the filter frequency response, which is usually the case). Frequency response is covered in Chapter 2.

Loop Rise Time

The loop rise time, τ_r, is defined as the 10 to 90% AGC response time to step changes in input power. The rise time is dependent on the nonlinear characteristics of the detector; thus the equations presented here (see Appendix E for the derivation) are valid only for small input steps (less than ±2 dBm). In Chapter 2 methods to improve this limitation are discussed.

The rise times for the integrator and low-pass-filter loops are the same, and may be given as

$$\tau_{r,SL} = \frac{9.14\ RC}{XA_{\Delta}A_{\epsilon}e_N} \tag{1-57}$$

$$\tau_{r,Lin} = \frac{18.3\ RC}{XA_{\Delta}A_{\epsilon}e_N} \tag{1-58}$$

As can be seen, the loop rise time is also dependent on $XA_{\Delta}A_{\epsilon}e_N$. However, the rise time can be calculated independently in terms of R and C.

The rise times given in Equations 1-57 and 1-58 are for CW inputs. What is the rise time if the loop is for pulses rather than CW? Chapter 2 will discuss a pulse AGC loop; however, the effect a pulse AGC loop has on τ_r is quite easy to determine. Figure 7 illustrates a basic pulse AGC loop, and the pertinent timing is shown in Figure 8.

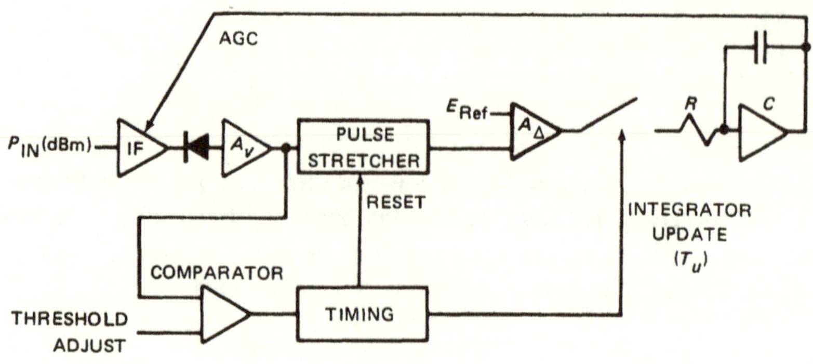

FIGURE 7. Basic Pulse AGC Block Diagram.

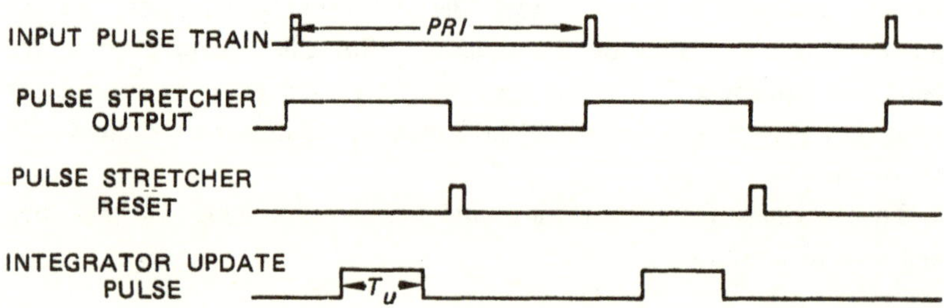

FIGURE 8. Basic Pulse AGC Timing.

The operation of the pulse AGC loop is simple; the input pulse is amplified, detected, stretched, and compared to the reference voltage, E_{Ref}. The amplified video also triggers a comparator (threshold) that initiates the timing. The integrator update switch is closed for a given period, T_u, every pulse; thus the integrator is only allowed to update the AGC loop during T_u. The effective time constant (T_{pulse}) for the updated integrator is

$$T_{pulse} = RC\left(\frac{PRI}{T_u}\right) \tag{1-59}$$

The update duty cycle, D, may be given as

$$D = \frac{T_u}{PRI} \tag{1-60}$$

Thus Equation 1-57 may be written as

$$T_{pulse} = \frac{RC}{D} \tag{1-61}$$

All equations thus presented for loop rise time, τ_r, may now be given in more general terms as

$$\tau_{r,SL} = \frac{9.14\ RC}{XA_\Delta A_\epsilon e_N D} \tag{1-62}$$

$$\tau_{r,Lin} = \frac{18.3\ RC}{XA_\Delta A_\epsilon e_N D} \tag{1-63}$$

It is obvious from Equations 1-62 and 1-63 that the loop rise times will be increased for decreased duty cycle.

It should be noted that sampling the AGC loop makes it a sampled data system. This publication will assume that any input modulation frequency is much smaller (by at least 10) than the sampling rate, or PRF in pulsed AGC. If this condition is not met, our equations fall apart.

AUTOMATIC GAIN CONTROL DESIGN VERIFICATION

The preceding section presented the equations that characterize an AGC tracking loop. In this section we will verify these equations using a simple AGC tracking loop. Nonideal parameters will be discussed (i.e., the IF amplifier's variable gain slope, X, is not linear over the full AGC range), and practical design equations will be presented with these nonideal parameters in mind.

Square Law Detector Test Circuit

Figure 9 illustrates the circuit used to validate the equations already presented. The philosophy here is to analyze an existing AGC loop to verify the equations rather than verify by design (Chapter 2 presents the design for several practical AGC loops). The square law detector will be discussed first, then the linear.

Figure 10 illustrates the square law characteristics for the detector used (see Appendix B for a discussion of detector characteristics). A normalized video reference, e_N, of -1 volt will be used; thus the detector output, e_D, is

$$e_D = e_N / A_v \tag{1-64}$$

or, since $A_v = 240$,

$$e_D = 1/240 = 4.2 \text{ mV} \tag{1-65}$$

The detector input power required to give an e_D of 4.2 mV is, referring to Figure 10,

$$P_{PD}(\text{dBm}) \cong -19 \text{ dBm} \tag{1-66}$$

Detector D_2 is used as a temperature stabilizing element to minimize the DC offset effects of D_1 (with no IF input, R_a is adjusted until $e_N = 0$ volt). Resistor R_b is adjusted to give the desired normalized video voltage, e_N (-1 volt). The -6-dBm power splitter enables reading the output power variation. The IF output power is

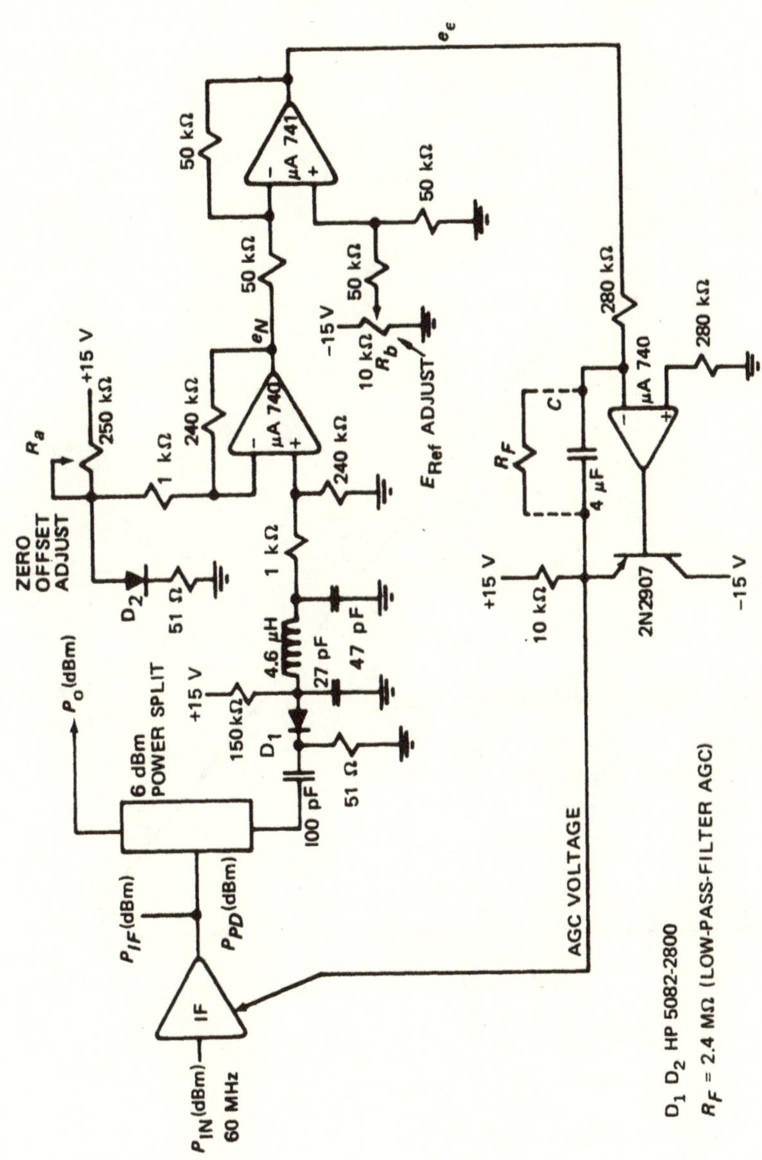

FIGURE 9. **Square Law Detector AGC Test Circuit.**

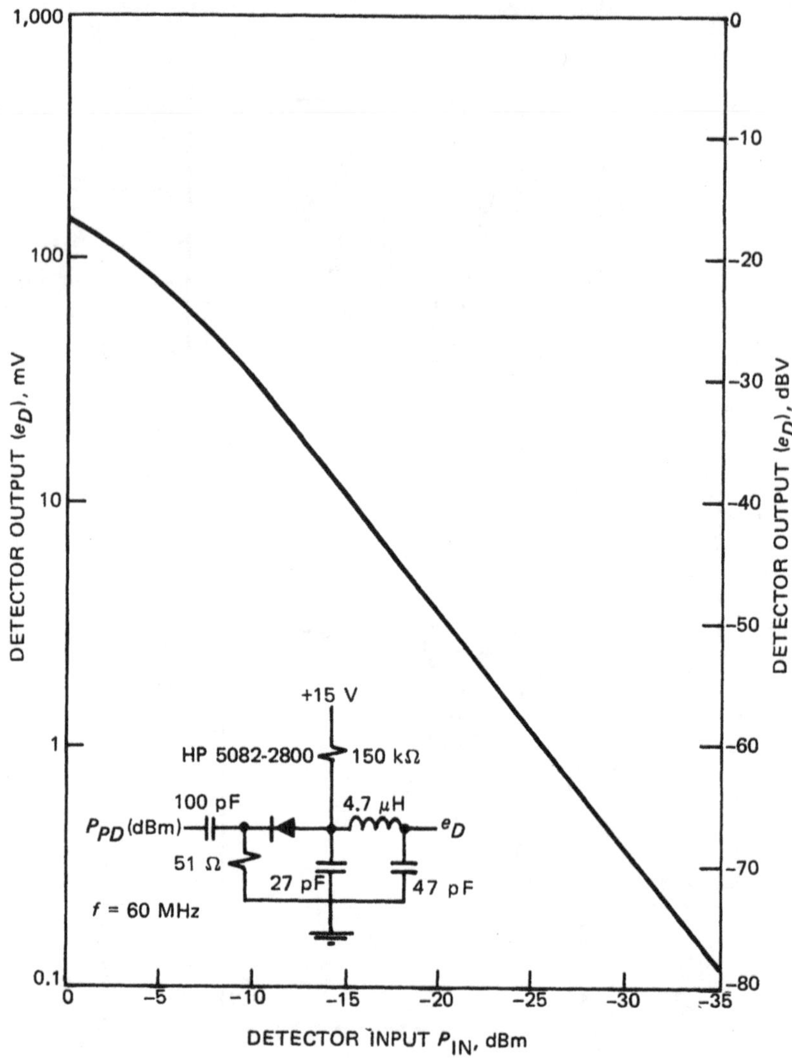

FIGURE 10. Detector Output Versus Input Power (HP 5082-2800).

$$P_{IF}\,(\text{dBm}) = P_{PD}\,(\text{dBm}) + 6\ \text{dBm} \tag{1-67}$$

where both outputs of the power splitter are the same (P_o(dBm) = P_{PD}(dBm)). Resistor R_F is used for the low-pass-filter AGC.

Figure 11 illustrates the variable gain characteristics of the IF amplifier. It will be noticed that the variable gain slope varies with AGC voltage (this is typical of many commercial variable gain IF amplifiers). The variation in X with AGC voltage and gain is given in Table 1. As can be seen, there is more than a three-to-one variation in X. Thus the loop gain, static regulation (for the low-pass-filter AGC), and rise time of the loop will be a function of AGC voltage; thus the input power, P_{in}(dBm).

The static regulation for the low-pass-filter AGC was presented as (Equation 1-5)

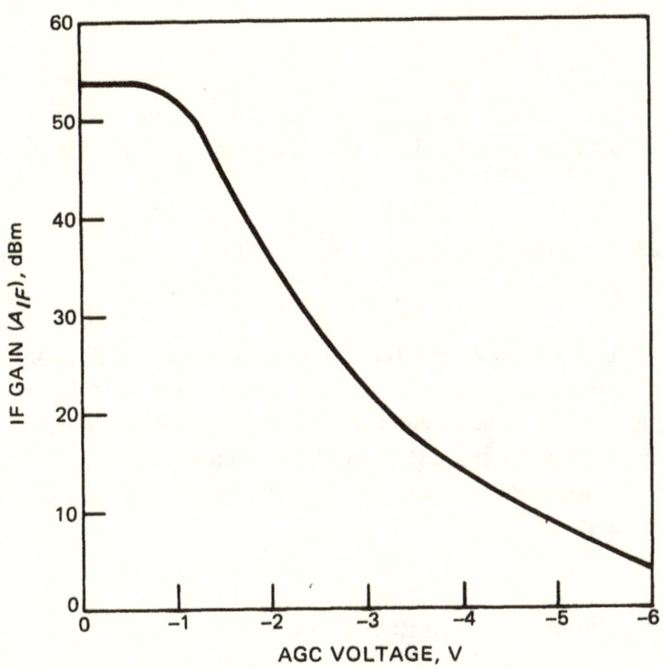

FIGURE 11. IF Amplifier Variable Gain Characteristics.

TABLE 1. IF Amplifier Variable Gain Slope.

AGC voltage, V	Gain, dBm	X, dBm/V
−1.75	40	18.2
−2.85	30	13.9
−3.25	20	9.1
−4.6	10	5.5

$$\Delta P_{o,SL}(\text{dBm}) = 10 \log \left[\frac{\Delta P_{\text{in}}(\text{dBm})}{X A_{\Delta} A_{\epsilon} e_N} + 1 \right] \tag{1-68}$$

which assumes a linear gain slope, X. Obviously the gain slope varies (Figure 11 and Table 1), and Equation 1-68 is not valid. This problem is easily corrected, however, by noting that

$$\Delta AGC = \frac{\Delta P_{\text{in}}(\text{dBm})}{X} \tag{1-69}$$

where ΔAGC is the total change in AGC voltage for the total wanted input dynamic range, $\Delta P_{\text{in}}(\text{dBm})$. Equation 1-69 may now be written as

$$\Delta P_{o,SL}(\text{dBm}) = 10 \log \left[\frac{\Delta AGC}{A_{\Delta} A_{\epsilon} e_N} + 1 \right] \tag{1-70}$$

which defines the static regulation for a practical low-pass-filter AGC loop using a square law detector (the linear detector loop will be presented shortly). The static regulation for the integrator loop will still be near zero, due to the large static (DC) gain of the integrator.

The parameters for the low-pass-filter AGC loop may now be given as (see Figure 9)

$A_{PD}(\text{dBm}) = -6; P_{PD}(\text{dBm}) = -19; P_{IF}(\text{dBm}) = -13$

$e_N = -1$ (adjusted at -3 volts AGC); $A_v = 240$

$A_{\Delta} = 1; A_{\epsilon} = 8.57; RC = 9.6$

The input was varied from -70 to -10 dBm ($\Delta P_{in}(\text{dBm}) = 60$), and the AGC voltage varied from -1.01 to -6.87 volts ($\Delta AGC = 5.86$ volts), as shown in Figure 12. The output power varied from -22.2 dBm at an AGC voltage of -1.01 volts to -19.3 dBm at an AGC voltage of 6.87 volts ($\Delta P_o(\text{dBm}) = 2.9$ dBm), as shown in Figure 13. Thus the change in output power, $\Delta P_o(\text{dBm})$, is

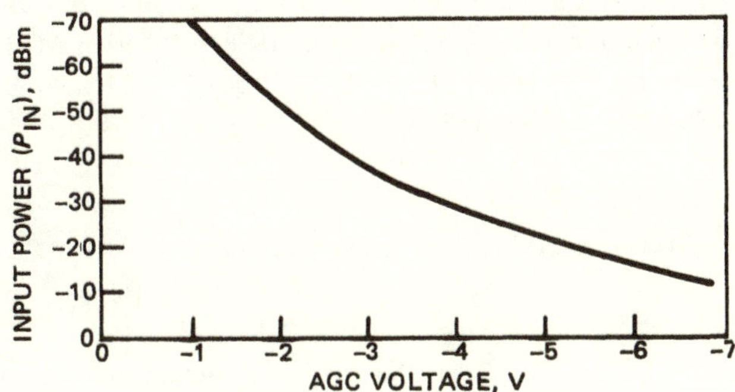

FIGURE 12. Input Power Versus AGC Voltage (Low-Pass Filter).

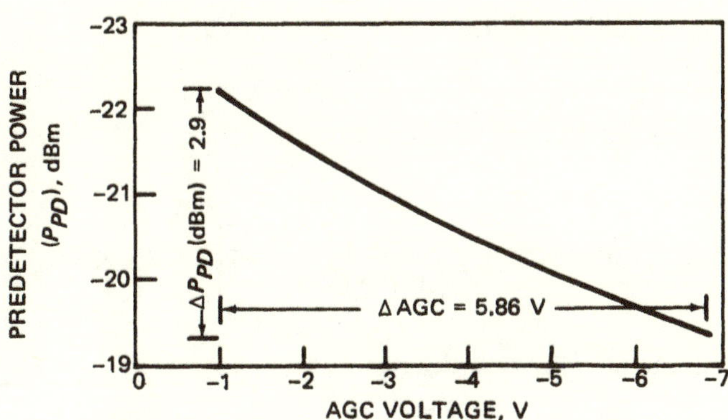

FIGURE 13. Detector Input Power Versus AGC Voltage (Low-Pass Filter).

$$\Delta P_{o,SL}(\text{dBm}) = 2.9 \tag{1-71}$$

The predicted change in output power, using Equation 1-70, is

$$\Delta P_{o,SL}(\text{dBm}) = 10 \log \left[\frac{5.86}{(1)\,(8.57)\,(1)} + 1 \right] = 2.26 \text{ dBm} \tag{1-72}$$

and is in excellent agreement with the measured value of 2.9 dBm. The normalized video voltage, e_N, varied from -0.78 volt for an AGC voltage of -1.01 volts to -1.42 volts at an AGC voltage of -6.87 volts. Thus the measured change in e_N, Δe_N(dBV), is

$$\Delta e_{N,SL}(\text{dBV}) = 20 \log (1.42/0.78) = 5.2 \text{ dBV} \tag{1-73}$$

The predicted value is twice the change in output power, since the change in e_D (thus e_N) in decibels with reference to 1 volt is twice the change in input power in decibels with reference to 1 milliwatt for a square law detector (see Appendix B). Thus,

$$\Delta e_{N,SL}(\text{dBV}) = 20 \log \left(\frac{\Delta AGC}{A_\Delta A_\epsilon e_N} + 1 \right) \tag{1-74}$$

or

$$\Delta e_{N,SL}(\text{dBV}) = 20 \log \left[\frac{5.86}{(1)\,(8.57)\,(1)} + 1 \right] = 4.5 \text{ dBV} \tag{1-75}$$

which also is in excellent agreement with the measured value of 5.2 dBV.

The value for e_N was changed to -2 volts by adjusting R_b, and the measured difference in ΔP_o(dBm) was

$$\Delta P_{o,SL}(\text{dBm}) = 1.23 \text{ dBm} \tag{1-76}$$

The predicted value, using Equation 1-70, is

$$\Delta P_{o,SL}(\text{dBm}) = 10 \log \left[\frac{5.86}{(1)\,(8.57)\,(2)} + 1 \right] = 1.28 \text{ dBm} \tag{1-77}$$

which again is in excellent agreement with the measured result. The measured change in e_N, $\Delta e_{N,SL}$(dBV), was 2.15 dBV. The predicted change is in excellent agreement with the measured value, as shown below.

$$\Delta e_{N,SL}(\text{dBV}) = 20 \log \left[\frac{5.86}{(1)\,(8.57)\,(2)} + 1 \right] = 2.55 \text{ dBV} \qquad (1\text{-}78)$$

The loop rise time was measured for input variations of ±1 dBm, ±5 dBm, and ±10 dBm at three input levels. Table 2 summarizes the results.

The equations predicting the loop rise time are valid only for small input deviations as discussed in the preceding section and Appendix E. The rise time for the low-pass-filter loop was given as (Equation 1-57)

$$\tau_{r,SL} = \frac{9.14\, RC}{X A_\Delta A_\epsilon e_N} \qquad (1\text{-}79)$$

TABLE 2. Loop Rise Time Results.

P_{in}, dBm	AGC, V	X, dBm/V	Loop rise time, τ_r, s					
			−1 dBm	+1 dBm	−5 dBm	+5 dBm	−10 dBm	+10 dBm
−53	−1.93	15	0.7	0.8	0.85	0.6	1	0.5
−37	−3.16	9.1	1	0.95	1.1	0.8	1.25	0.7
−26	−4.51	5.5	1.25	1.25	1.5	1.1	1.7	0.9

The predicted values are compared to the measured in Table 3 (±1 dBm), and as can be seen there is good agreement.

The deviation in loop rise time at large input variations (Table 2) is due to the nonlinear behavior of the detector (Appendix D). Two methods to minimize this effect will be given in Chapter 2.

The loop gain was measured as illustrated in Figure 6. The modulating frequency was well below the low-pass-filter bandwidth.

$$f_{3\text{dBV}}(LPF) \cong \frac{1}{2\pi RC} \qquad (1\text{-}80)$$

TABLE 3. Measured and Predicted Loop Rise Times.

P_{in}, dBm	X, dBm/V	τ_r (measured), s	τ_r (predicted—Eq. 1-79), s
−53	15	0.8	0.68
−37	9.1	1	1.13
−26	5.5	1.25	1.86

TABLE 4. Low-Pass-Filter Loop Gains.

P_{in}, dBm	AGC, V	X, dBm/V	LG
−50	−1.95	15	25.7
−35	−3.19	9.1	15
−25	−4.53	5.5	12.8

(The open loop frequency response for this circuit, C removed, is in excess of 3 kilohertz; thus the $R_F C$ filter will determine the bandwidth for frequencies up to about 1 kilohertz.)

The measured loop gains are listed in Table 4. The predicted loop gain is (Equation 1-36)

$$LG_{SL} = 0.23\, XA_{\Delta}A_{\epsilon}e_N \tag{1-81}$$

Table 5 compares the measured and predicted loop gains, which can be seen to be very close.

Figure 9 (with R_F removed) illustrates the basic integrator AGC loop using a square law detector. Figure 14 illustrates the input power versus AGC voltage for $e_N = -1$ volt. The measured change in the output power, ΔP_0(dBm), was 0.12 dBm inputs from −65 to −10 dBm. The predicted change, from Equation 1-22, is 0 dBm, which is in excellent agreement with the measured result.

Table 6 summarizes the loop rise time results.

The predicted rise time, again valid only for small input variations (Equation 1-79), is given in Table 7, and it agrees favorably with the measured results. The value for e_N was doubled, and the loop rise time was halved, as predicted by Equation 1-79.

TABLE 5. Comparison of Measured and Predicted Loop Gains.

P_{in}, dBm	X, dBm/V	LG (measured)	LG (predicted)
−50	15	25.7 (28.2 dBV)	29.6 (29.4 dBV)
−35	9.1	15 (23.5 dBV)	17.9 (25.1 dBV)
−25	5.5	12.8 (22.1 dBV)	10.8 (20.7 dBV)

TABLE 6. Integrator AGC Loop Rise Time Results.

P_{in}, dBm	AGC, V	X, dBm/V	Loop rise time, τ_r, s					
			−1 dBm	+1 dBm	−5 dBm	+5 dBm	−10 dBm	+10 dBm
−50	1.95	15	0.58	0.55	0.68	0.4	0.7	0.4
−35	3.19	9.1	0.95	0.95	1.1	0.9	1.2	0.6
−25	4.53	5.5	1.5	1.4	1.8	1.2	1.9	0.9

TABLE 7. Measured and Predicted Loop Rise Times.

P_{in}, dBm	X, dBm/V	τ_r (measured), s	τ_r (predicted), s
−50	15	0.58	0.68
−35	9.1	0.95	1.12
−25	5.5	1.5	1.86

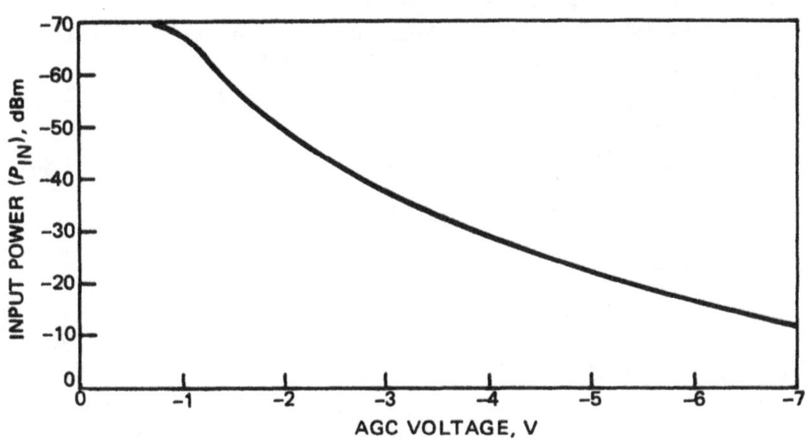

FIGURE 14. Input Power Versus AGC Voltage (Integrator, Square Law Detector).

The loop gain is frequency-sensitive, as discussed earlier.

$$LG_{SL} = \frac{0.036\ XA_{\Delta}A_{\epsilon}e_N}{fRC} \tag{1-82}$$

The measured loop gain for an input of −35 dBm (X = 9.1 dBm/V) was 0.03 with an input frequency of 10 hertz. The predicted value is

$$LG_{SL} = \frac{(0.036)\,(9.1)\,(1)\,(1)\,(1)}{(10)\,(280 \times 10^3)\,(4 \times 10^{-6})} = 0.0293 \tag{1-83}$$

which is in excellent agreement with the predicted results.

AGC loops employing linear detectors will now be covered, and the pertinent equations will be shown to be valid.

Linear Detector Test Circuit

Figure 15 illustrates the circuit used to verify the linear detector AGC equations. The same variable gain amplifier used for the square law detector test circuit is used, as is the detector. Figure 16 illustrates the linear characteristics for the HP 5082-2800 detector. As can be seen, this detector has linear qualities above 0 dBm.

A normalized video reference, e_N, of −1 volt is used; the detector output, e_D, is

$$e_D = e_N/A_v \tag{1-84}$$

A_v = 3.3; thus,

$$e_D = 1/3.3 = 0.3\ \text{V} \tag{1-85}$$

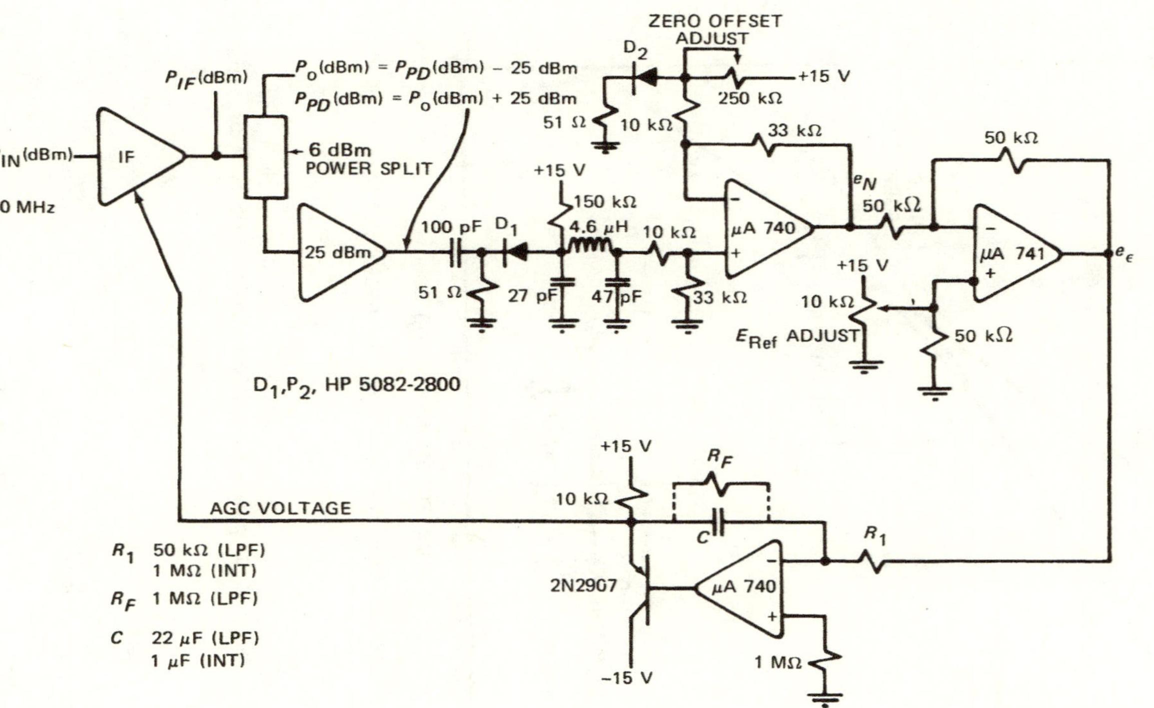

FIGURE 15. Linear Detector AGC Test Circuit.

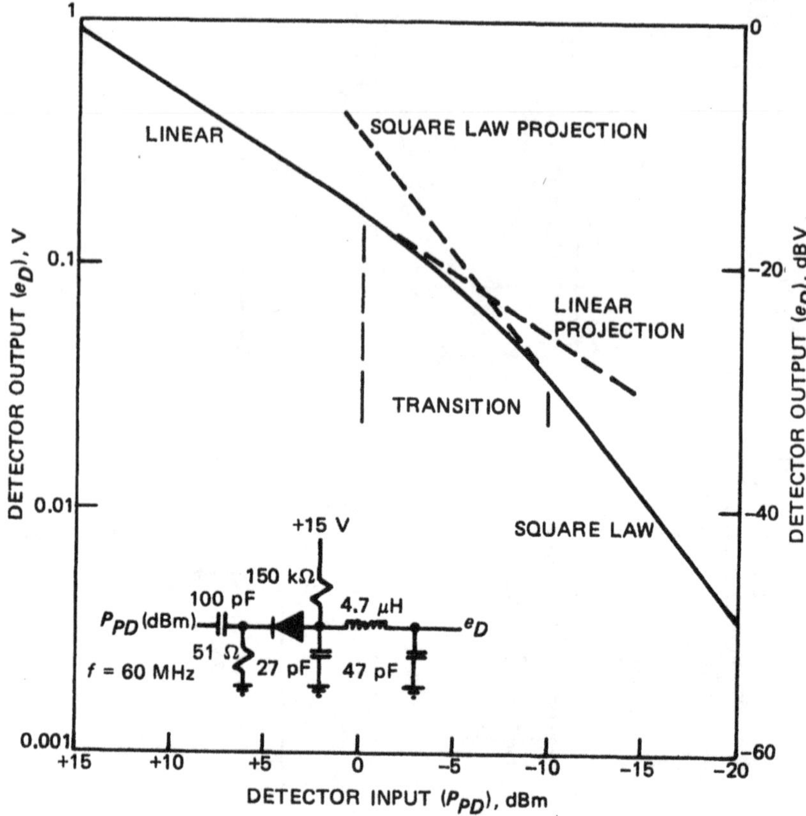

FIGURE 16. Detector Output Versus Input Power (HP 5082-2800).

The necessary input power to satisfy Equation 1-85 is 5 dBm (Figure 16). Thus,

$$P_{PD}(\text{dBm}) = 5 \text{ dBm} \tag{1-86}$$

and, referring to Figure 15,

$$P_{IF}(\text{dBm}) = 5 - 25 + 6 = -14 \text{ dBm} \tag{1-87}$$

The low-pass-filter AGC will be discussed first ($A_\nu = 3.3$, $A_\Delta = 1$, $A_\epsilon = 20$, $e_N = -1$ volt). The static regulation may be given as

$$\Delta P_{o,Lin}(\text{dBm}) = 20 \log \left(\frac{\Delta AGC}{A_{\Delta} A_{\epsilon} e_N} + 1 \right) \tag{1-88}$$

The input power was varied from −69 to 0 dBm (ΔP_{in}(dBm) = 69), and the AGC voltage varied from −0.82 to −7.03 volts (ΔAGC = 6.21 volts), as shown in Figure 17. The output power from the 6-dBm power splitter varied from −19.6 to −17.36 dBm (ΔP_o(dBm) = 2.24) over the AGC range, as shown in Figure 18. The nonlinearities for inputs larger than

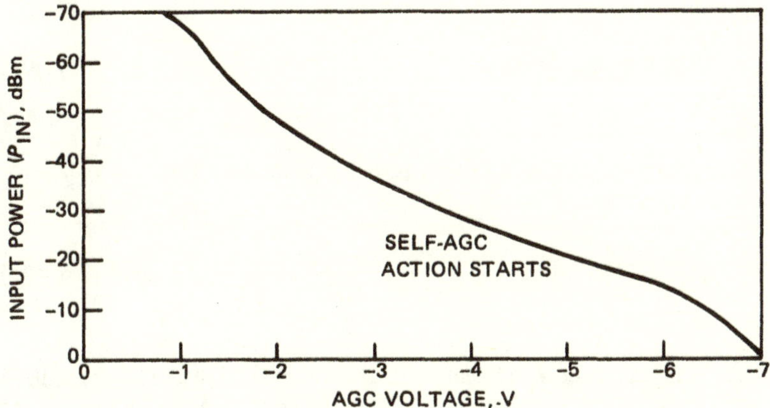

FIGURE 17. Input Power Versus AGC Voltage (LPF, Linear Detector).

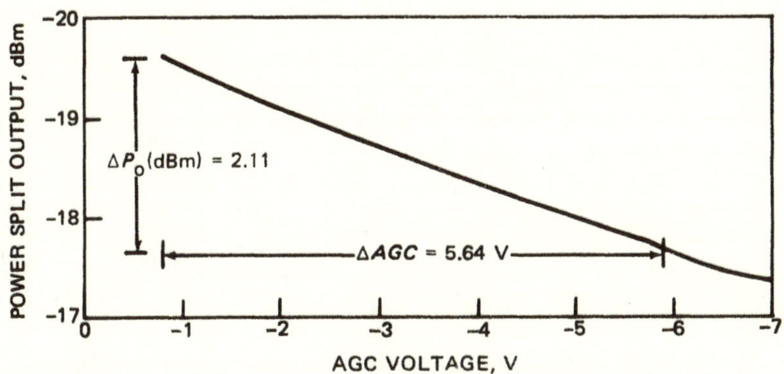

FIGURE 18. Power Splitter Output Power Versus AGC Voltage (LPF, Linear Detector).

−10 dBm are due to the self-AGC effects of the input signal on the gain (the *signal* is large enough to have a controlling effect on gain). This condition must be avoided for linear operation. To avoid any effects of self-AGC, only inputs from −69 to −10 dBm will be used. Using Figures 17 and 18,

$$\Delta P_{\text{in}}(\text{dBm}) = -59 \text{ dBm} \tag{1-89}$$

$$\Delta AGC = 5.64 \text{ V} \tag{1-90}$$

$$\Delta P_{\text{o}}(\text{dBm}) = 2.11 \text{ dBm} \tag{1-91}$$

The predicted change in output power is (Equation 1-88)

$$\Delta P_{\text{o},Lin}(\text{dBm}) = 20 \log \left[\frac{5.64}{(1)\,(1)\,(20)} + 1 \right] = 2.16 \text{ dBm} \tag{1-92}$$

which is in excellent agreement with the measured value. The normalized video output, e_N, varied from −0.895 volt ($AGC = -0.82$ volt) to −1.143 volts ($AGC = -6.46$ volts). Thus the change in e_N is (in dBV)

$$\Delta e_N(\text{dBV}) = 20 \log (1.143/0.895) = 2.12 \text{ dBV} \tag{1-93}$$

This is the same as the change in output power, which is to be expected for a linear detector. Thus

$$\Delta e_{N,Lin}(\text{dBV}) = 20 \log \left(\frac{\Delta AGC}{A_\Delta A_\epsilon e_N} + 1 \right) = \Delta P_{\text{o}}(\text{dBm}) \tag{1-94}$$

The loop rise time was measured for input variations of ±1 dBm ($AGC = 3.1$; $X = 9.1$ dBm/V) and was 1.7 seconds. The predicted loop rise time is (Equation 1-58)

$$\tau_{r,Lin} = \frac{18.3\ R_F C}{X A_\Delta A_\epsilon e_N} \tag{1-95}$$

or

$$\tau_{r,Lin} = \frac{18.3\ (1 \times 10^6)\ (22 \times 10^{-6})}{(9.1)\ (1)\ (20)\ (1)} = 2.21 \text{ seconds} \tag{1-96}$$

which is in good agreement with the measured value.

The loop gain was measured to be 22 ($AGC = -3.1$, $X = 9.1$ dBm/V). The predicted loop gain is (Equation 1-39)

$$LG_{Lin} = 0.12\ X A_\Delta A_\epsilon e_N \tag{1-97}$$

or

$$LG_{Lin} = 0.12\ (9.1)\ (1)\ (20)\ (1) = 21.8 \tag{1-98}$$

which is in excellent agreement with the measured value.

The linear detector, integrator AGC loop is illustrated in Figure 15, with R_F removed, $C = 1\mu$F, and $R_1 = 1$ MΩ. Figure 19 illustrates the input power versus AGC voltage. This curve also deviates from the expected at inputs larger than −10 dBm due to the self-AGC effects.

The power output from the 6-dBm power splitter varied from −18.3 dBm (P_{PD}(dBm) = 6.7 dBm) for −65 dBm input power to −18.53 dBm (P_{PD}(dBm) = 6.47 dBm) for −10 dBm input power. Thus the output power changed:

$$\Delta P_o(\text{dBm}) = 6.7 - 6.47 = 0.23 \text{ dBm} \tag{1-99}$$

which is very close to the 0-dBm change predicted. The normalized video voltage, e_N, varied from −1.042 to −1.018 volts over the same input power range. Thus,

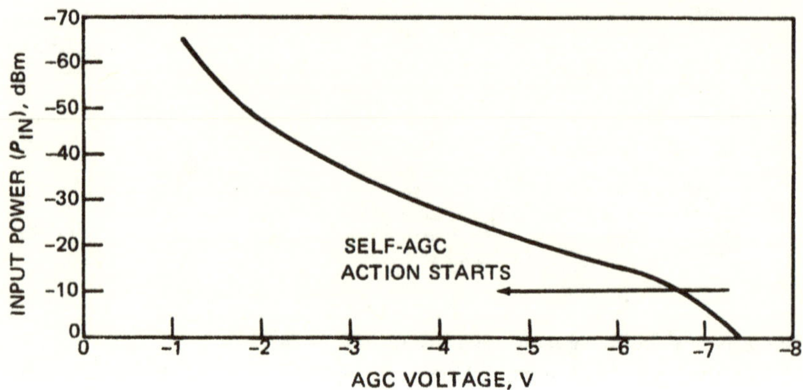

FIGURE 19. Input Power Versus AGC Voltage (Integrator, Linear Detector).

$$\Delta e_N(\text{dBV}) = 20 \log \frac{1.042}{1.018} = 0.2 \text{ dBV} \tag{1-100}$$

which is similar to the change in output power, as is to be expected with the linear detector.

The loop rise time measured for an AGC voltage of −3.1 (X = 9.1 dBm/V) at ±1 dBm input power deviation was 1.8 seconds. The predicted value (Equation 1-95) is

$$\tau_r = \frac{18.3\ (1 \times 10^6)\ (1 \times 10^{-6})}{(9.1)\ (1)\ (1)\ (1)} = 2.01 \text{ seconds} \tag{1-101}$$

and is in excellent agreement with the measured value.

The loop gain is 0.0147 for a modulation frequency of 10 hertz. The predicted value (Equation 1-56) is

$$LG_{Lin} = \frac{0.018\ XA_{\Delta}A_e e_N}{fRC} \tag{1-102}$$

or

$$LG_{Lin} = \frac{0.018\ (9.1)\ (1)\ (1)\ (1)}{10\ (1 \times 10^6)\ (1 \times 10^{-6})} = 0.0164 \tag{1-103}$$

which is also in excellent agreement with the measured value.

A series switch was placed between the differencing amplifier, A_Δ, and the integrator to verify Equations 1-62 and 1-63. A pulse repetition frequency of 1 kilohertz and update time, T_u, of 500 microseconds were used. The duty cycle is thus

$$D = \frac{T_u}{PRI} = T_u\,(PRF) \tag{1-104}$$

or

$$D = (500 \times 10^{-6})\,(1 \times 10^3) = 0.5 \tag{1-105}$$

The expected loop rise times are, from Equations 1-62 and 1-63,

$$\tau_{r,SL} = \frac{9.14\,RC}{XA_\Delta A_\epsilon e_N D} \tag{1-106}$$

$$\tau_{r,Lin} = \frac{18.3\,RC}{XA_\Delta A_\epsilon e_N D} \tag{1-107}$$

or the loop rise times should double for the condition defined. Measured results show exact agreement, and Equations 1-62 and 1-63 are valid.

The frequency response for the low-pass-filter loop gain (loop gain decreased 3 dBV from its low frequency value) may be given as

$$f_{3\text{dBV}}(LG) = \frac{D}{2\pi RC} = \frac{0.159\,D}{RC} \tag{1-108}$$

It should be noted that a holding capacitor (sample-hold circuit) is needed to hold the error voltage, e_ϵ, for the low-pass-filter AGC in pulse-AGC applications.

This chapter has presented the basic theory and pertinent operation equations for AGC loops, and the theory has been verified with a practical AGC test circuit.

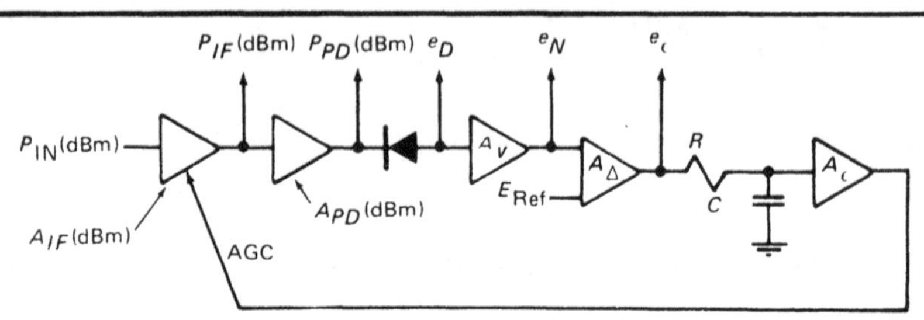

n = 1 (linear detector) $\qquad$ n = 2 (square law detector)

Static Regulation

$$\Delta P_o(\text{dBm}) = \frac{20}{n} \log\left(\frac{\Delta AGC}{A_\Delta A_\epsilon e_N} + 1\right) \qquad A_\Delta A_\epsilon e_N = \frac{\Delta AGC}{10^{\frac{n \Delta P_o(\text{dBm})}{20}} - 1}$$

$$\Delta P_o(\text{dBm}) = \frac{20}{n} \log\left[\frac{a(\Delta AGC)\, 10^{\frac{-P_{PD}(\text{dBm})n}{20}}}{K A_v A_\Delta A_\epsilon} + 1\right]$$

$$A_v A_\Delta A_\epsilon = \frac{a(\Delta AGC)\, 10^{\frac{-P_{PD}(\text{dBm})n}{20}}}{K\left[10^{\frac{n \Delta P_o(\text{dBm})}{20}} - 1\right]}$$

a = 1.59 (linear detector) $\qquad$ a = 2.5 (square law detector)

$$\Delta e_N(\text{dBV}) = n \Delta P_o(\text{dBm}) \qquad A_\Delta A_\epsilon e_N = \frac{\Delta AGC}{10^{\frac{\Delta e_N(\text{dBV})}{20}} - 1}$$

FIGURE 20. Low-Pass-Filter AGC Design Equation Summary.

$$\Delta P_o(\text{dBm}) = \frac{20}{n} \log\left[\frac{0.12\, nX\Delta AGC}{LG} + 1\right] \qquad \Delta e_N(\text{dBV}) = 20 \log\left[\frac{0.12\, nX\Delta AGC}{LG} + 1\right]$$

Dynamic Regulation

$$LG = 0.12\, nXA_\Delta A_\epsilon e_N \qquad A_\Delta A_\epsilon e_N = \frac{LG}{0.12\, nX}$$

$$LG = bKXA_V A_\Delta A_\epsilon\, 10^{\frac{P_{PD}(\text{dBm})n}{20}} \qquad A_V A_\Delta A_\epsilon = \frac{(LG)\, 10^{\frac{-P_{PD}(\text{dBm})n}{20}}}{bXK}$$

$b = 0.0756$ (linear detector) $\qquad$ $b = 0.096$ (square law detector)

$$LG = 0.12\, nX\left[\frac{\Delta AGC}{10^{\frac{n\Delta P_o(\text{dBm})}{20}} - 1}\right] \qquad LG = 0.12\, nX\left[\frac{\Delta AGC}{10^{\frac{\Delta e_N(\text{dBV})}{20}} - 1}\right]$$

Loop Rise Time

$$\tau_r = \frac{18.3\, RC}{nXA_\Delta A_\epsilon e_N D} \qquad \tau_r = \frac{d(RC)\, 10^{\frac{-P_{PD}(\text{dBm})n}{20}}}{XKA_V A_\Delta A_\epsilon D}$$

$$f_{3\text{dBV}}(LG) = \frac{0.159\, D}{RC} \qquad (LG)(\tau_r) = \frac{2.2\, RC}{D} \qquad \tau_r = \frac{0.35}{[LG][f_{3\text{dBV}}(LG)]}$$

$d = 29$ (linear detector) $\qquad$ $d = 22.9$ (square law detector)

D = LPF update duty cycle

FIGURE 20. (Contd.)

The designer now has the information necessary to design an AGC loop to meet individual needs. If static regulation is of paramount importance, the integrator AGC loop can be used, adjusting R and C for the wanted response time. If input modulation is of interest, the low-pass-filter AGC loop can be used, defining R and C to give the wanted loop gain frequency response.

Figures 20 and 21 summarize the equations presented in this chapter, and will be used in the applications presented in the next chapter.

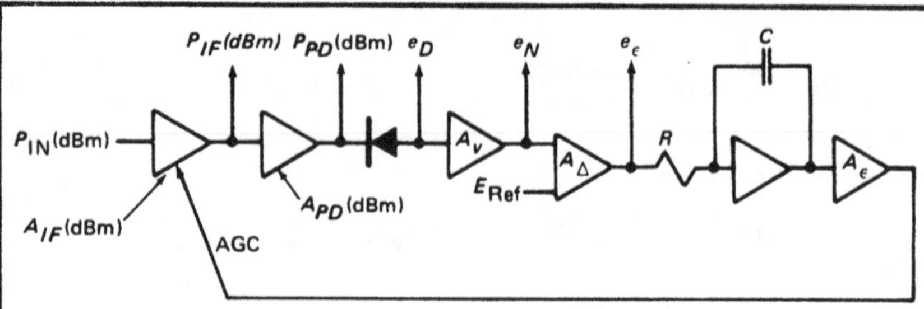

$n = 1$ (linear detector) $\quad$ $n = 2$ (square law detector)

Static Regulation

$\Delta P_o(\text{dBm}) = 0 \qquad \Delta e_N(\text{dBV}) \cong 0$

Dynamic Regulation

$$LG = \frac{0.018\, nXA_\Delta A_\epsilon e_N}{fRC} \qquad A_\Delta A_\epsilon e_N = \frac{(LG) fRC}{0.018\, nX}$$

$$LG = \frac{aXKA_V A_\Delta A_\epsilon\, 10^{\frac{P_{PD}(\text{dBm})n}{20}}}{fRC} \qquad A_V A_\Delta A_\epsilon = \frac{(LG)(fRC)\, 10^{\frac{-P_{PD}(\text{dBm})n}{20}}}{aXK}$$

$a = 0.0113$ (linear detector) $\quad$ $a = 0.0144$ (square law detector)

f = input modulation frequency

Loop Rise Time

$$\tau_r = \frac{18.3\, RC}{nXA_\Delta A_\epsilon e_N D} \qquad \tau_r = \frac{b(RC)\, 10^{\frac{-P_{PD}(\text{dBm})n}{20}}}{XKA_V A_\Delta A_\epsilon D}$$

$$(LG)(\tau_r) = \frac{2.2\, RC}{D}$$

$b = 29$ (linear detector) $\quad$ $b = 22.9$ (square law detector)

D = integrator update duty cycle

FIGURE 21. Integrator AGC Design Equation Summary.

2

Applications

Chapter 1 presented and verified the pertinent equations necessary to characterize and design AGC loops. The design of an individual loop will depend on the critical design parameter, static regulation, input modulation reduction, etc. This chapter will present several examples to demonstrate the necessary techniques. A method to overcome the nonlinear rise time with input steps greater than ±2 dB will also be presented.

CONTINUOUS WAVE LOW-PASS-FILTER AGC LOOP

A 30-megahertz AGC loop is wanted that will operate over a bandwidth of at least 50 hertz and have an input modulation reduction of 0.01 (the output modulation is 0.01 times the input modulation, or the output increases 0.01% for each 1% increase on the input). The other design parameters are

$$P_{in,min}(\text{dBm}) = -60;\ P_{in,max}(\text{dBm}) = -10;\ P_o(\text{dBm}) = -15$$

A back diode detector, GE BD-2, will be used because of its low (near zero) offset voltage, which minimizes the need for detector temperature compensation.[2] The BD-2 has a diode constant of

[2] H. A. Watson. *Microwave Semiconductor Devices and Their Circuit Applications.* New York, McGraw-Hill Book Co., 1969.

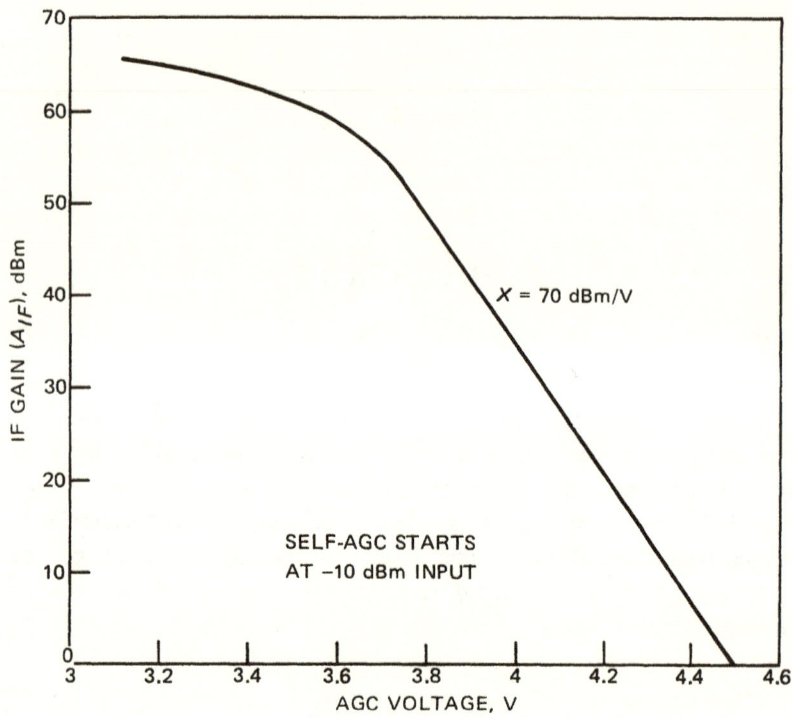

FIGURE 22. IF Gain Versus AGC Voltage (30-MHz IF Amplifier).

$$K_{SL} = 1.2 \tag{2-1}$$

and −15 dBm is well in the square law region; thus,

$$P_o(\text{dBm}) = P_{PD}(\text{dBm}) \tag{2-2}$$

Figure 22 illustrates the variable gain characteristic of the IF amplifier used. As can be seen,

$$X = 70 \text{ dBm/V} \tag{2-3}$$

and the AGC characteristics are quite linear over most of its usable range.

A 6-dBm power splitter will be used at the IF amplifier's output as shown in Figure 23. Thus,

$$P_{IF}(\text{dBm}) = -15 + 6 = -9 \text{ dBm} \tag{2-4}$$

and is well within the output capabilities of the amplifier used.

To conserve components, A_Δ and A_ϵ will be combined as shown in Figure 23. The value for A_ν and $A_\Delta A_\epsilon$ may be given as (Figure 20)

$$A_\nu A_\Delta A_\epsilon = \frac{10.4\,(LG)\,10^{\frac{-P_{PD}(\text{dBm})n}{20}}}{XK_{SL}} \tag{2-5}$$

where, from Chapter 1,

$$LG \cong \frac{1}{IMR} = 100 \tag{2-6}$$

Now

$$A_\nu A_\Delta A_\epsilon = \frac{(10.4)\,(100)\,10^{\frac{-(-15)2}{20}}}{(70)\,(1.2)} = 392 \tag{2-7}$$

A video gain of 100 will be used, leaving

$$A_\Delta A_\epsilon = \frac{392}{100} = 3.92 \tag{2-8}$$

The complete AGC loop is illustrated in Figure 23. R_a was set for a P_o(dBm) of -15 dBm ($E_{\text{Ref}} = -0.277$) for an AGC voltage of 4.12 (near the center of the linear AGC characteristic). Figure 24 illustrates the loop

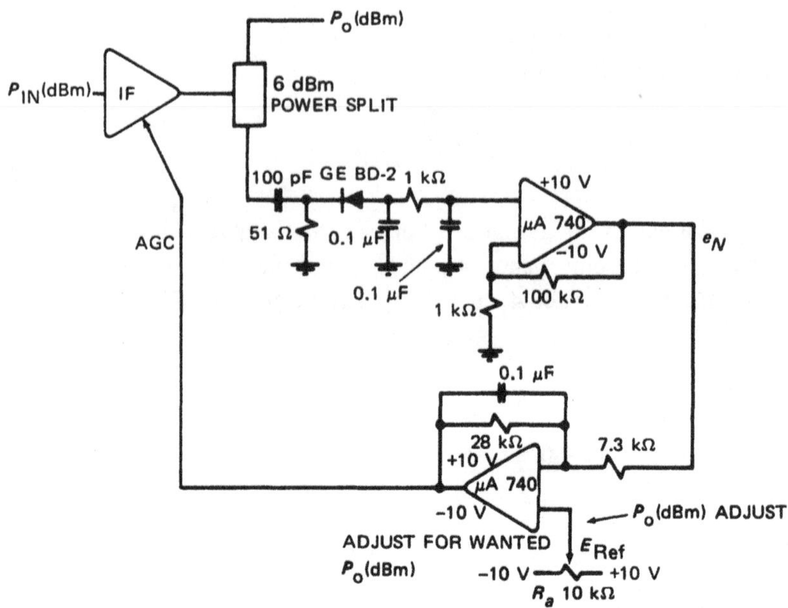

FIGURE 23. 30-MHz AGC Loop.

AGC characteristics. The loop gain was measured and found to be 90, which is very close to the designed value. Figure 25 illustrates the static regulation properties, and, as can be seen,

$$\Delta P_o(\text{dBm}) = 0.5 \tag{2-9}$$

for $P_{in,min}(\text{dBm}) = -60$ and $P_{in,max}(\text{dBm}) = -10$. The change in AGC voltage over this range is

$$\Delta AGC = 0.71 \text{ V} \tag{2-10}$$

and the predicted static regulation may now be found;

$$\Delta P_o(\text{dBm}) = 10 \log \left[\frac{2.5 \, \Delta AGC \, 10^{\frac{-P_{PD}(\text{dBm})n}{20}}}{K_{SL} A_\nu A_\Delta A_\epsilon} + 1 \right] \tag{2-11}$$

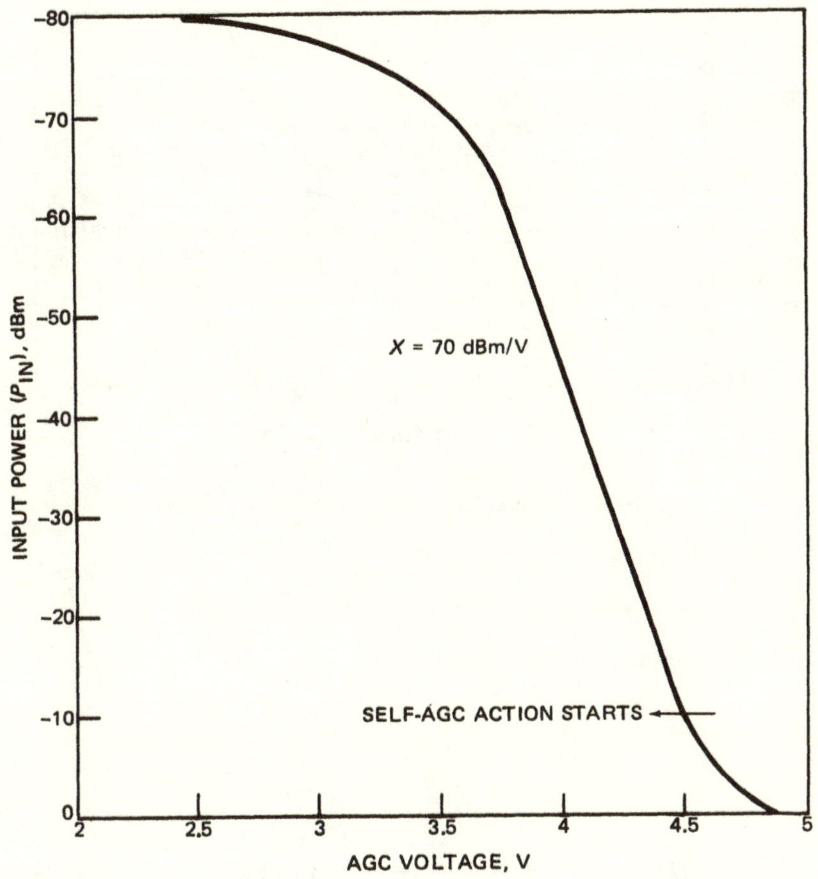

FIGURE 24. AGC Voltage Versus Input Power (30-MHz Continuous Wave AGC).

or

$$\Delta P_o(\text{dBm}) = 10 \log \left[\frac{(2.5)\,(0.71)\,10^{\frac{-(-15)2}{20}}}{(1.2)\,(392)} + 1 \right] = 0.49 \text{ dBm} \qquad (2\text{-}12)$$

This is in excellent agreement with the measured results.

The loop gain bandwidth, for $C = 0.1\ \mu\text{F}$, is

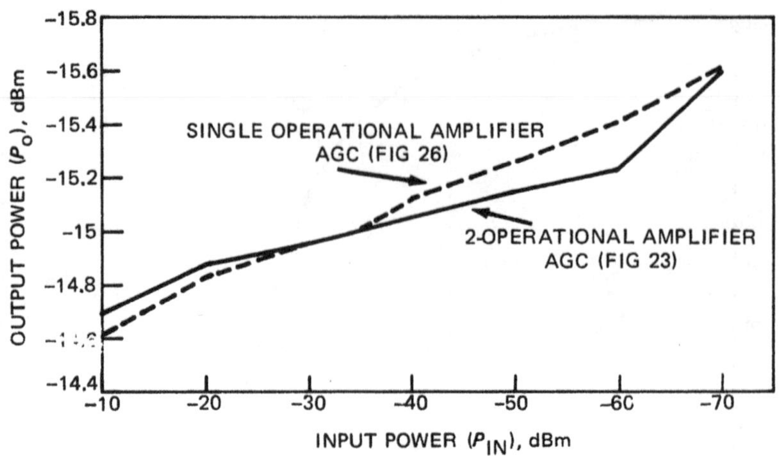

FIGURE 25. Output Power Deviation Versus Input Power.

$$f_{3\,\mathrm{dB}}(LG) = \frac{1}{2\pi(RC)} \tag{2-13}$$

or

$$f_{3\,\mathrm{dB}}(LG) = \frac{1}{(6.28)\,(28\ \mathrm{k}\Omega)\ 0.1\ \mu\mathrm{F}} = 56.8\ \mathrm{Hz} \tag{2-14}$$

and the measured bandwidth was 60 hertz. The predicted loop rise time is (Figure 20)

$$\tau_r = \frac{0.35}{[LG]\,[f_{3\,\mathrm{dBV}}(LG)]} = \frac{2.2\ RC}{LG} \tag{2-15}$$

or

$$\tau_r = 61\ \mu\mathrm{s} \tag{2-16}$$

At this point the reader may ask, "Can the two operational amplifiers of Figure 23 be combined?" The answer is, of course, yes. Figure 26 illustrates the ultimate in AGC loop simplicity. All calculations performed for the two-operational-amplifier case apply to Figure 26 as well (C is decreased to 0.01 microfarad due to the increase of R to 392 kilohms).

The loop gain was measured at 90. The static regulation is illustrated in Figure 25 (the AGC characteristics are the same as in the two-operational-amplifier case) and is very close to the predicted. The measured loop gain bandwidth is 42 hertz, which compares well with the predicted value of 40 hertz. If this loop is to have a minimum bandwidth of 50 hertz, decrease C.

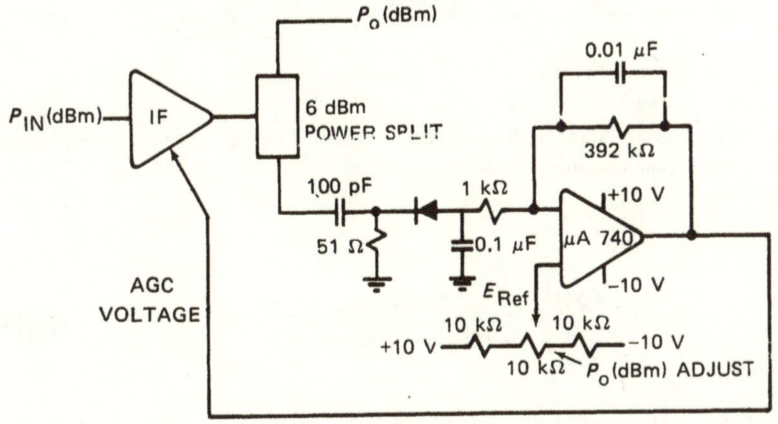

FIGURE 26. Single Operational Amplifier 30-MHz AGC Loop.

SCANNING BEAM AGC LOOPS

There is much interest lately in scanning beam radars. The beam may be scanned either electrically or mechanically, but the end result is the same: the return is a series of pulses amplitude-modulated by the beam pattern, as shown in Figure 27a. Figure 27b illustrates the scan timing and typical returns. The number of return pulses (p) depends on the pulse repetition frequency (PRF), scan frequency (start scan to start scan), live time, scan width (θ), and bandwidth (BW),

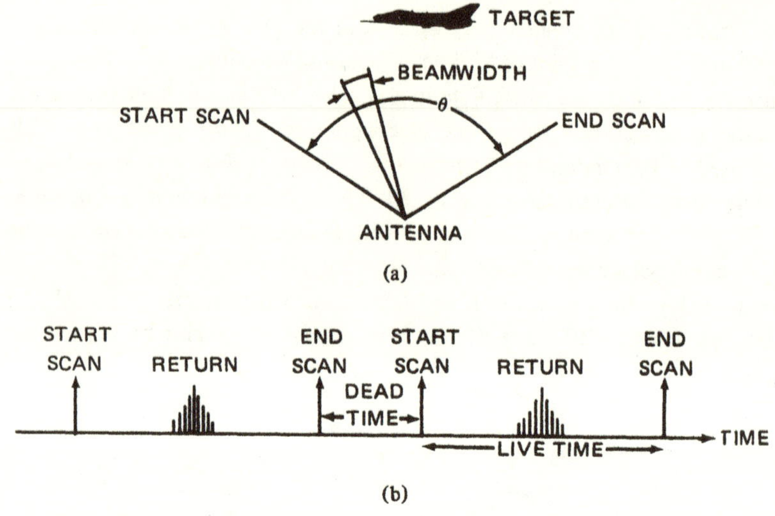

FIGURE 27. Scanning Beam Radar. (a) Basic scanning pattern, (b) basic scan timing and video return.

$$p = PRF \left(\frac{BW}{\theta}\right) \left(\frac{\%\ \text{live time}}{100\ \text{scan frequency}}\right) \tag{2-17}$$

One method of AGC for scanning beam radars is to normalize the maximum video return as shown in Figure 28. This technique normalizes the scan-to-scan video return. The peak detector holds the peak voltage, as shown, and the AGC loop normalizes this voltage during the integrator update time. The loop rise time may be easily calculated as (Figure 21)

$$\tau_r = \frac{18.3\ RC}{nXA_\Delta A_\epsilon e_N D} \tag{2-18}$$

where

n = 1 (linear detector), 2 (square law detector)

D = integrator update duty cycle

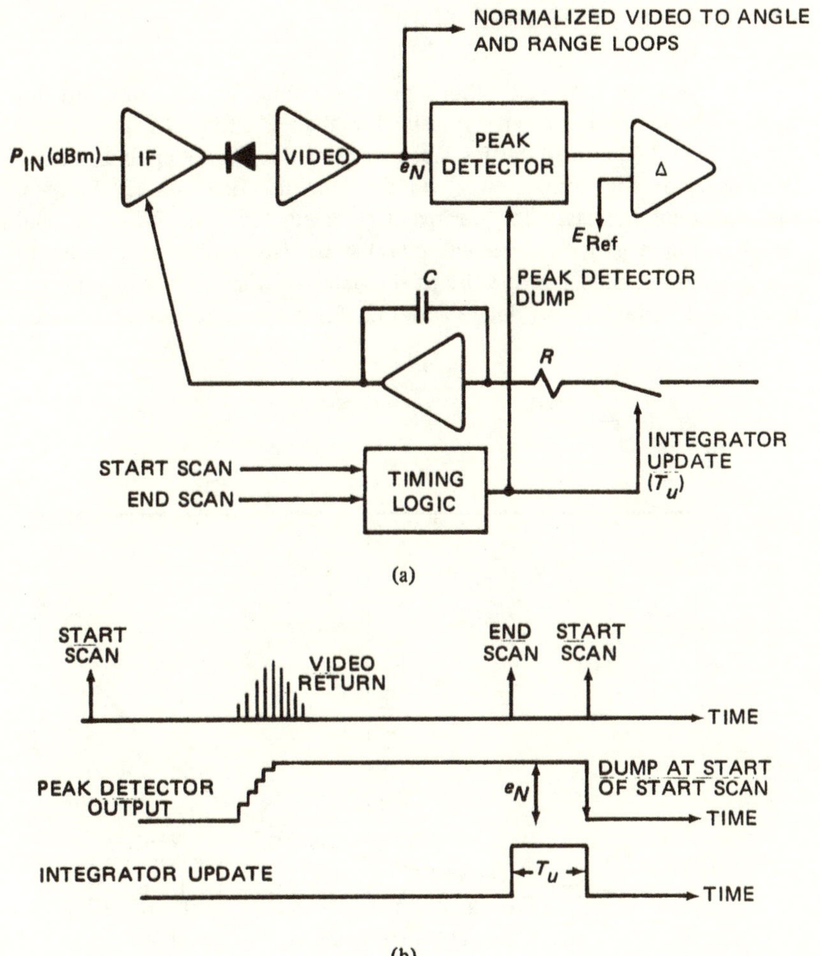

FIGURE 28. AGC Technique for a Scanning Beam Radar. (a) AGC loop, (b) timing.

D may be given as (Figure 28)

$$D = \frac{\text{dead time}}{\text{scan time}} \tag{2-19}$$

or

$$D = (\text{dead time})\,(\text{scan frequency}) \tag{2-20}$$

OPTIMIZING LOOP RISE TIMES

The loop rise time calculations presented thus far are valid only for small steps in input power, typically less than ±2 dBm. Tables 2 and 6 (pages 29 and 31) illustrate the dependence of the input step and polarity (increasing or decreasing power) on the loop rise time. As can be seen, the rise time increases for positive (increased power) input steps, and decreases for negative (decreased power) steps. This nonlinearity is due to the inherent nonlinear properties of the detector. Figure 29 illustrates the normalized instantaneous detector output for a square law detector, using

$$e_D = 10^{\frac{\Delta P_{in}(\text{dBm})}{10}} \tag{2-21}$$

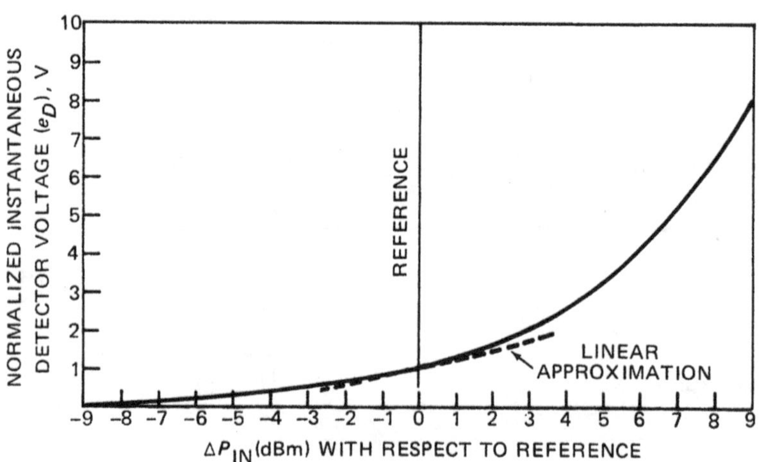

FIGURE 29. Instantaneous Detector Voltage Versus ΔP_{in}(dBm).

As can be seen, e_D is very nonlinear for ΔP_{in}(dBm) greater than ±2 dBm. Thus the AGC integrator, or low-pass-filter input will be very large for $+\Delta P_{in}$(dBm) and very small for $-\Delta P_{in}$(dBm) (e.g., for ΔP_{in}(dBm) = +5, $\Delta e_D \cong 3.2 - 1 = 2.2$, and for ΔP_{in}(dBm) = −5, $\Delta e_D = 1 - 0.3 = 0.7$). Thus the rise times for large $+\Delta P_{in}$(dBm) should be expected to be shorter than for large $-\Delta P_{in}$(dBm).

Certain AGC loops may require a more constant loop rise time with large values of $\pm\Delta P_{in}$(dBm). Two solutions to this problem are apparent from the previous discussion: (1) have a shorter RC time constant for $-\Delta P_{in}$(dBm) steps, or (2) linearize e_D versus ΔP_{in}(dBm). These two techniques will now be presented.

Variable Time Constant

Figure 30 illustrates the variable time constant circuit. This circuit replaces the differencing circuit and integrator of Figure 9. The diode, D_1, blocks positive voltages (or $+\Delta P_{in}$(dBm)), and the loop behaves as it normally does; however, D_1 turns on for $-\Delta P_{in}$(dBm), thus decreasing the

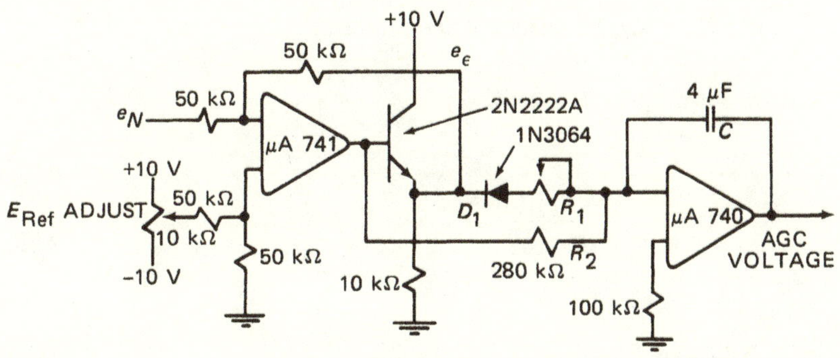

FIGURE 30. Variable Time Constant Circuit.

integrator's RC time constant (R_1 and R_2 are now in parallel). The transistor compensates for the 0.6-volt DC diode drop. R_1 may now be adjusted to give equal rise times for $\pm\Delta P_{in}$(dBm). This idea works for ΔP_{in}(dBm) $\cong \pm 5$ dBm, but falls apart after that. A much better solution would be to linearize e_N with respect to ΔP_{in}(dBm).

Linearized Time Constant

A logarithmic amplifier preceding the differencing amplifier (Figure 31) will linearize e_N with respect to ΔP_{in}(dBm). A logarithmic amplifier has the following characteristics (Appendix F briefly discusses logarithmic amplification).

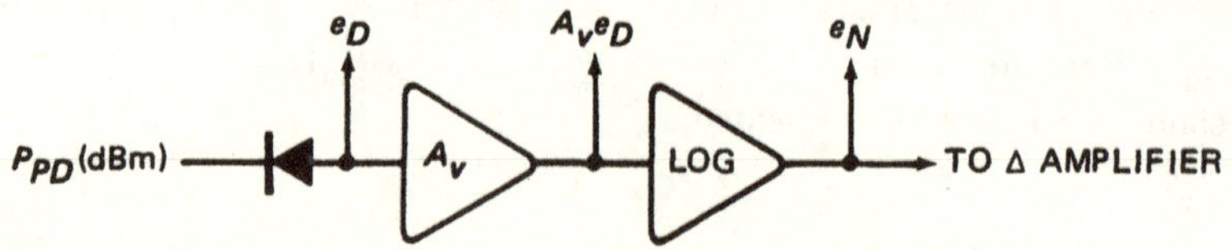

FIGURE 31. AGC Loop Employing Logarithmic Amplifier.

$$e_N = K_1 \log K_2 (A_v e_D) \tag{2-22}$$

where

K_1, K_2 = logarithmic amplifier constants

Now, for a square law detector,

$$e_D = 0.4\,K_{SL}\,10^{\frac{P_{PD}(\mathrm{dBm})}{10}} \tag{2-23}$$

and substituting Equation 2-23 into 2-22,

$$e_N = K_1 \log K_2 A_v \left[0.4\,K_{SL}\,10^{\frac{P_{PD}(\mathrm{dBm})}{10}}\right] \tag{2-24}$$

or

$$e_N = \frac{K_1 P_{PD}(\mathrm{dBm})}{10} + K_1 \log (K_2 A_v\ 0.4\,K_{SL}) \tag{2-25}$$

Taking the derivative of Equation 2-25 with respect to P_{PD}(dBm),

$$\frac{\Delta e_N}{\Delta P_{PD}(\mathrm{dBm})} = \frac{K_1}{10} \quad (\mathrm{V/dBm}) \tag{2-26}$$

or, the logarithmic video output has a constant slope of $K_1/10$ V/dBm, which is the wanted result.

To calculate the static regulation, loop gain, and loop rise time, simply include the logarithmic amplifier in the calculations given in Appendixes C, D, and E. The derivation of these equations is left as an exercise for the reader. Figures 32 and 33 illustrate the pertinent parameters for the logarithmic amplifier AGC loop.

An AGC loop will now be designed using the logarithmic amplifier to optimize the loop rise time.

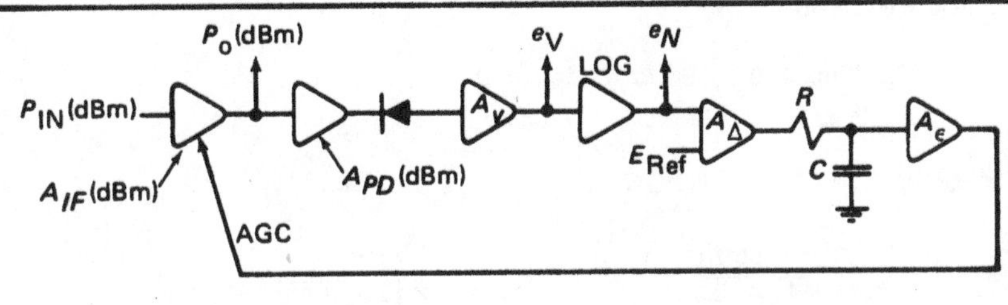

X = IF gain slope (dBm/V)

LS = logarithmic slope (v/dBV)

D = update duty cycle

Static Regulation

$$\Delta P_{o,SL}\,(\text{dBm}) = \frac{\Delta AGC}{2A_\Delta A_\epsilon (LS)} \qquad \Delta e_{N,SL}\,(\text{dBm}) = \frac{\Delta AGC}{A_\Delta A_\epsilon (LS)}$$

$$\Delta P_{o,Lin}\,(\text{dBm}) = \frac{\Delta AGC}{A_\Delta A_\epsilon (LS)} \qquad \Delta e_{N,Lin}\,(\text{dBm}) = \frac{\Delta AGC}{A_\Delta A_\epsilon (LS)}$$

Loop Rise Time (10 to 90%)

$$\tau_{r,SL} = \frac{1.05\,RC}{XA_\Delta A_\epsilon (LS)D} \qquad \tau_{r,Lin} = \frac{2.1\,RC}{XA_\Delta A_\epsilon (LS)D}$$

Loop Gain

$$LG_{SL} = 2\,XA_\Delta A_\epsilon (LS) \qquad LG_{Lin} = XA_\Delta A_\epsilon (LS)$$

$$f_{3dBV}(LG) = \frac{1}{2\pi RC} \qquad (LG)\,(\tau_r) = \frac{2.2\,RC}{D}$$

FIGURE 32. Low-Pass Filter and Log Video AGC Design Summary.

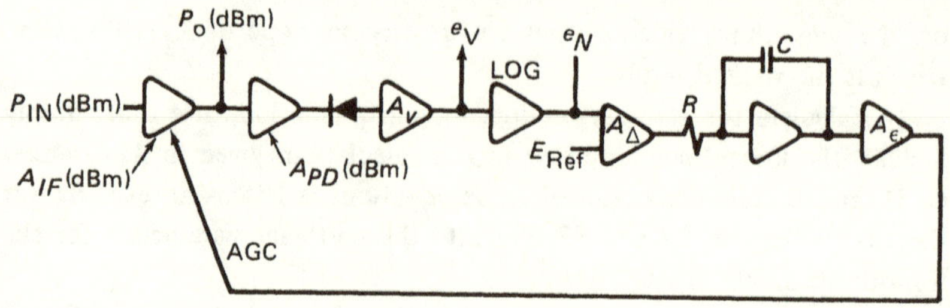

X = IF gain slope (dBm/V)

LS = logarithmic slope (V/dBm)

D = integrator update duty cycle

Static Regulation

$\Delta P_o(\text{dBm}) \cong 0 \qquad \Delta e_N(\text{dBm}) \cong 0$

Loop Rise Time (10 to 90%)

$$\tau_{r,SL} = \frac{1.05\, RC}{XA_\Delta A_\epsilon (LS)D} \qquad \tau_{r,Lin} = \frac{2.1\, RC}{XA_\Delta A_\epsilon (LS)D}$$

Loop Gain

$$LG_{SL} = \frac{0.315\, XA_\Delta A_\epsilon (LS)D}{fRC} \qquad LG_{Lin} = \frac{0.158\, XA_\Delta A_\epsilon (LS)D}{fRC}$$

$$(LG)\ (\tau_r) = \frac{0.331}{f}$$

f = input modulation frequency

FIGURE 33. Integrator and Log Video AGC Design Summary.

OPTIMIZED RISE TIME, AGC LOOP

Figure 34 illustrates the pulse AGC configuration to be discussed. The IF amplifier is a commercially available 60-megahertz unit. The AGC characteristics for this amplifier are illustrated in Figure 35, and as will be seen, X varies from 6.25 to 17.5 dBm/V over its useful range.

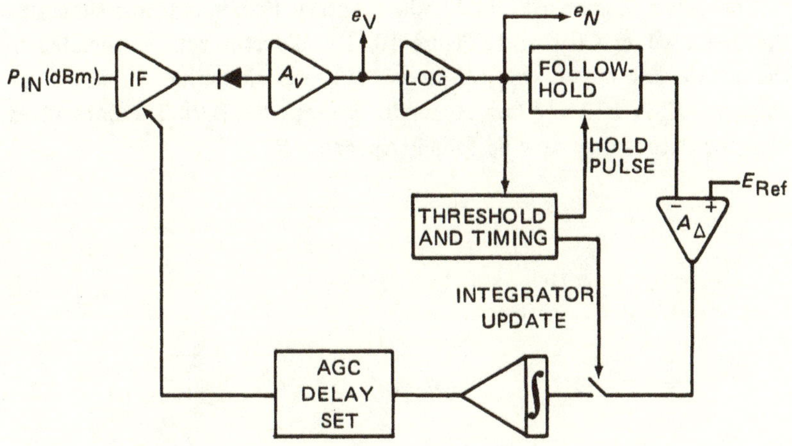

FIGURE 34. Linear Rise Time AGC Loop.

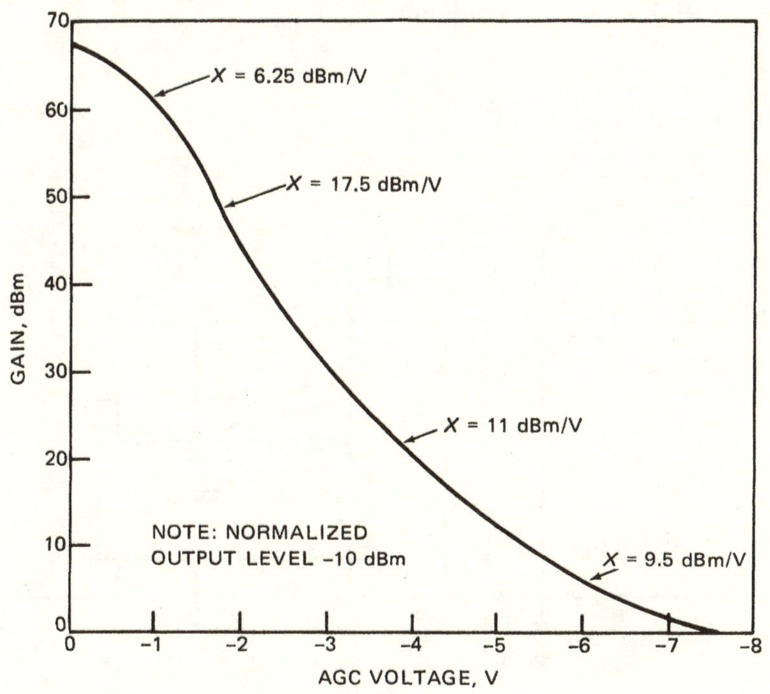

FIGURE 35. Variable Gain Characteristics for 60-MHz IF Amplifier.

The detector is an HP 5082-2800 Schottky barrier detector similar to the one used in Chapter 1, Figure 10. The detector will be operated in the square law region (P_{PD}(dBm) = −15), and the value for the diode constant, K_{SL}, is 0.879 (see Appendix B, Equation B-9). The normalized detector output may now be found (Appendix B).

$$e_D = 0.4\,K_{SL}\,10^{\frac{P_{PD}(\text{dBm})}{10}} \tag{2-27}$$

or

$$e_D = (0.4)\,(0.879)\,10^{\frac{-15}{10}} = 11 \text{ mV} \tag{2-28}$$

A 733 video amplifier with a gain of 10 (20 dBV) will be used; thus the video output, e_v, will be (Figure 36)

$$e_v = 10\,e_D = 110 \text{ mV} \tag{2-29}$$

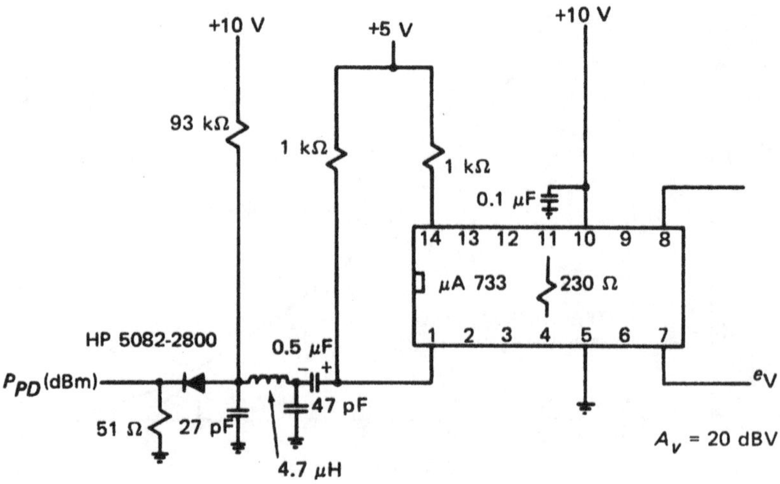

FIGURE 36. 60-MHz Pulse AGC Detector-Video Amplifier.

The normalized video voltage, e_N, is to be 2 volts, with an input of 110 millivolts. The logarithmic video amplifier is illustrated in Figure 37 and discussed more fully in the reference cited;[3] it has the following characteristics:

$$LS = \frac{I_T R_c}{40} \text{ V/dBV} \tag{2-30}$$

where

LS = logarithmic slope

I_T = 2 mA

R_c = 2 kΩ

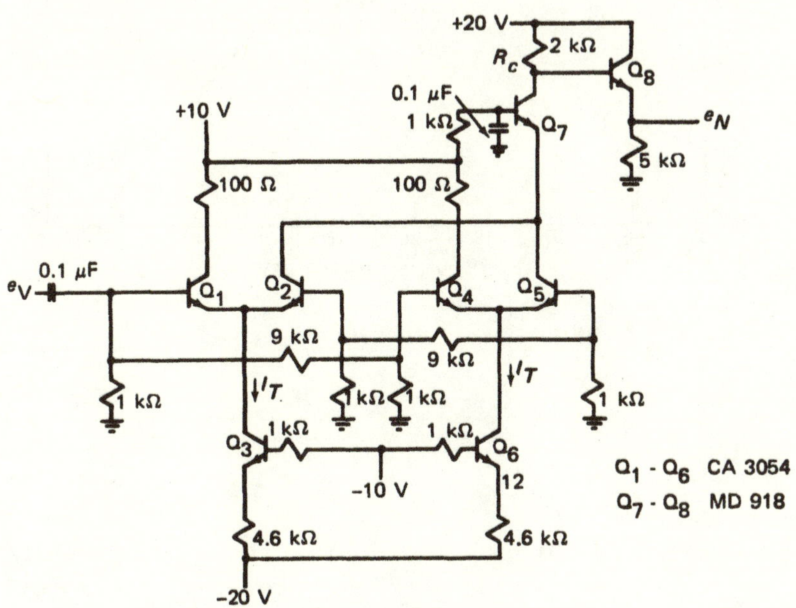

FIGURE 37. Pulse AGC Log Video Amplifier.

[3] R. S. Hughes. *Logarithmic Video Amplifiers.* Dedham, Mass., Artech House, Inc., 1971.

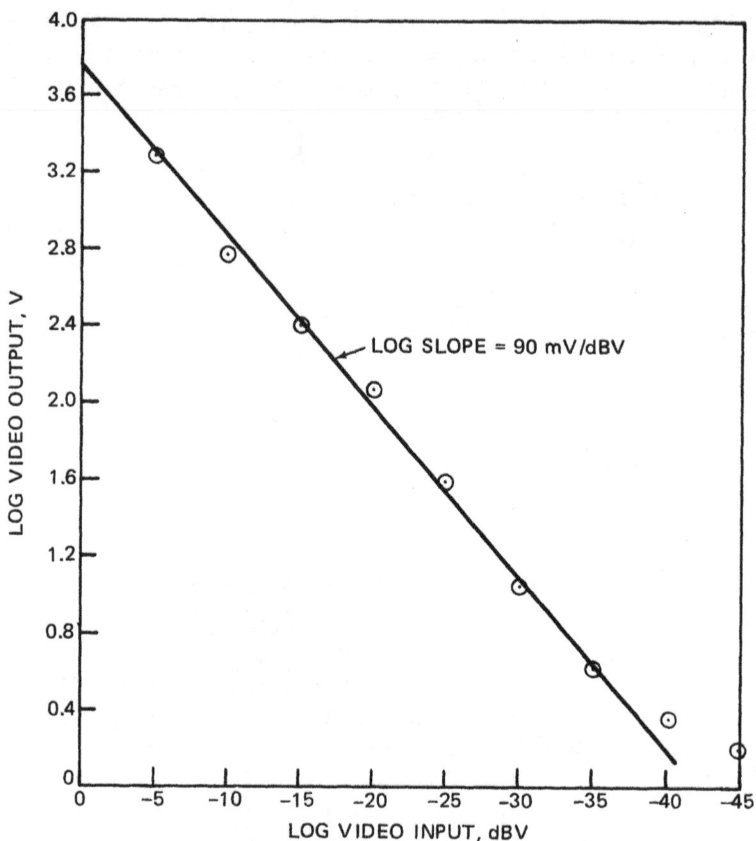

FIGURE 38. Log Video Output Versus Input.

or

$$LS = 100 \text{ mV/dBV} \tag{2-31}$$

Figure 38 illustrates the logarithmic response for the log video. The measured logarithmic scale factor is 90 mV/dBV, which is very close to the predicted value of 100 mV/dBV. It is apparent from the above figure that an input of −20 dBV (100 millivolts) will give the wanted output of 2 volts (the predicted video output for a detector input of −15 dBm is 110 millivolts). The instantaneous input dynamic range into the log video is ±20 dBV (for a normalized 2-volt output); this corresponds to a dynamic input power change, for a square law detector, of ±10 dBm. Thus this AGC loop will have an optimized loop rise time for input steps up to ±10 dBm. The loop will be adjusted by varying E_{Ref} until $e_N = 2$ volts.

Figure 39 illustrates the follow-hold amplifier. Transistors Q_1 and Q_2 form a series switch. The switch is normally closed (Q_1 and Q_2 saturated), but it opens when pin 11 on the μA-9017 (driver gate) is grounded. The capacitor holds this level until pin 11 goes to +5 volts. The gain of the follow-hold amplifier is +1.

Figure 40 illustrates the differencing amplifier-integrator combination. Switch S_1 is the integrator update switch, and S_2 is a DC reset switch to

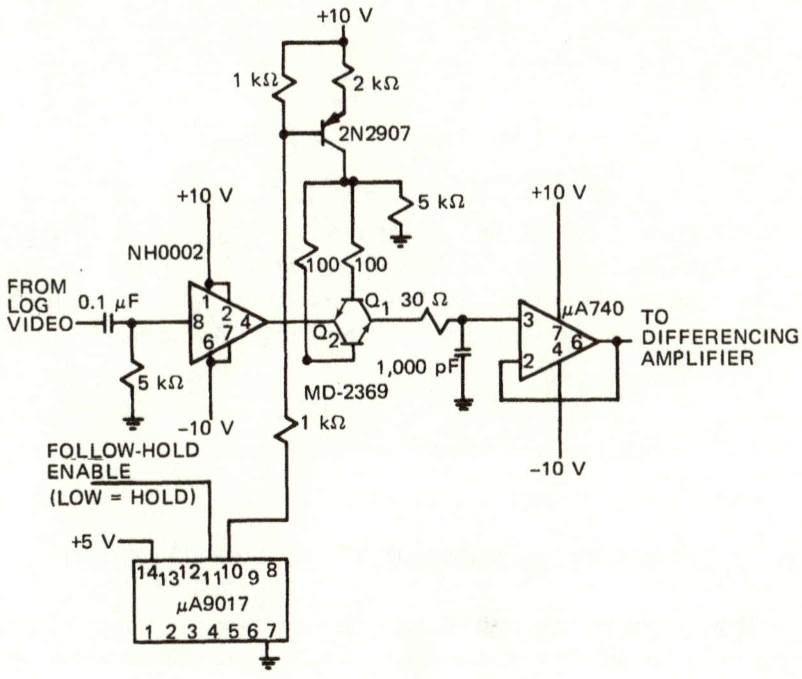

FIGURE 39. Follow-Hold Amplifier.

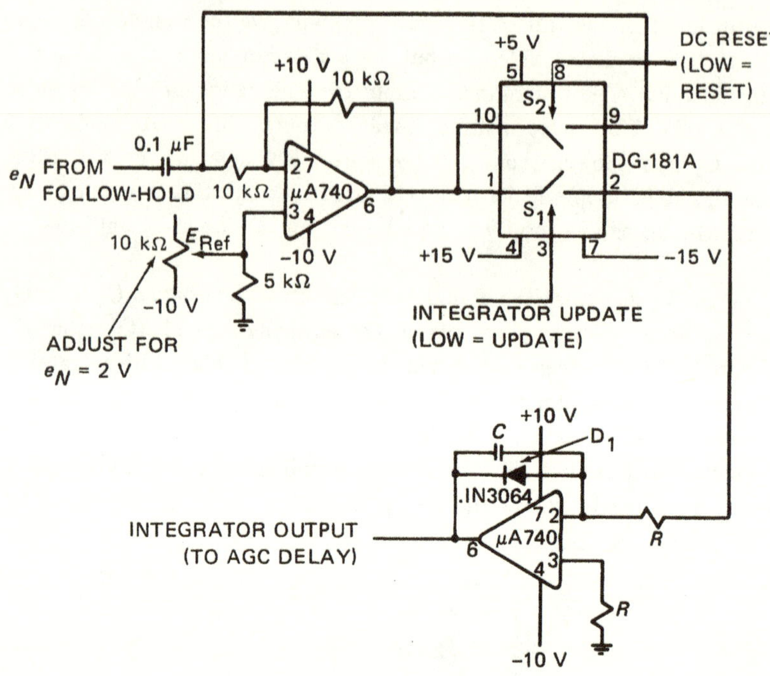

FIGURE 40. Differencing Amplifier-Integrator.

normalize the charge on the coupling capacitor (this increases the pulse repetition frequency rate that the loop can handle). The output of the differencing amplifier (with no pulse present) is E_{Ref}. The loop is normalized when the input pulse to the differencing amplifier, e_N, is equal to E_{Ref}:

$$e_N = E_{Ref}$$

where

e_N = normalized differencing amplifier input pulse (2 volts)

Now, if the IF input should increase after loop normalization, the output of the differencing amplifier will be a negative pulse (Figure 41a). This negative pulse will be gated by S_1, and the integrator output will increase until the pulse is again normalized (Figure 41b). The diode, D, across the integrating capacitor prevents the integrator from saturating.

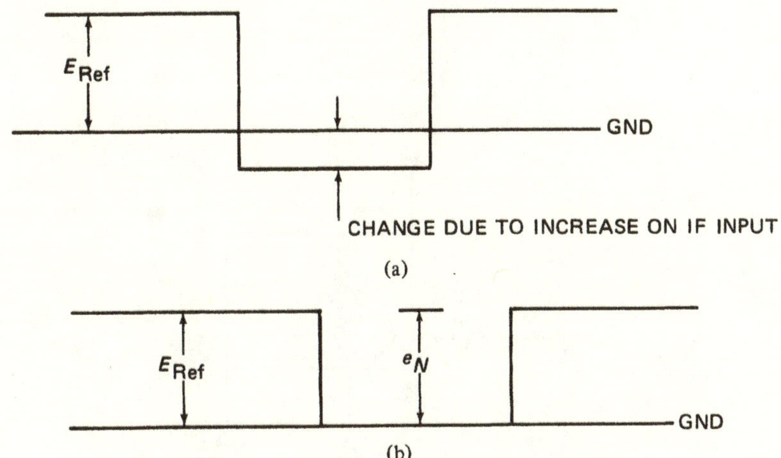

FIGURE 41. Differencing Amplifier Output. (a) Increased IF input, (b) normalized output.

Figure 42 illustrates the AGC delay (when AGC action starts) and AGC drive circuitry, and Figures 43 and 44 show the threshold and timing circuitry. The time delay between thresholding and the hold pulse is controlled by the 390 Ω resistor and the 300 pF capacitor (Figure 44), and was chosen to be 250 nanoseconds for this circuit. A hold pulse of 7 microseconds and an integrator update time of 5 microseconds was also used as shown in Figure 45. The DC reset switch is closed except during the 5-microsecond hold time.

It is sometimes necessary to design pulse AGC loops for a given number of pulses (i.e., the AGC loop must close in N pulses). This may easily be accomplished as follows. The loop rise time may be given as (Figure 33)

$$\tau_{r,SL} = \frac{1.05\ RC}{XA_{\Delta}A_{\epsilon}(LS)D} \tag{2-32}$$

D is the update duty cycle (Equation 1-60),

$$D = \frac{T_u}{PRI} \tag{2-33}$$

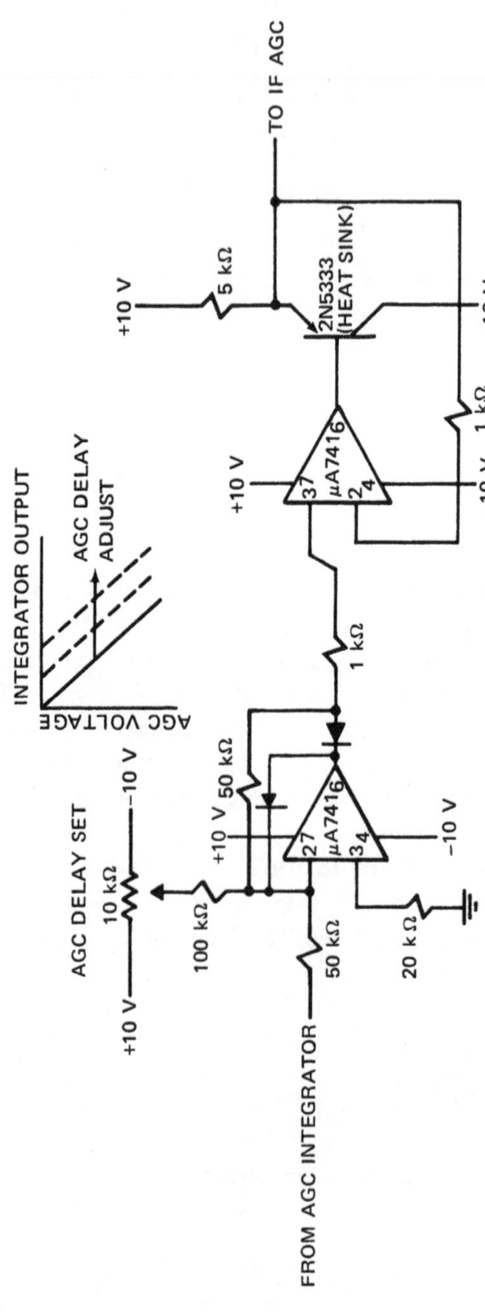

FIGURE 42. AGC Delay and AGC Drive.

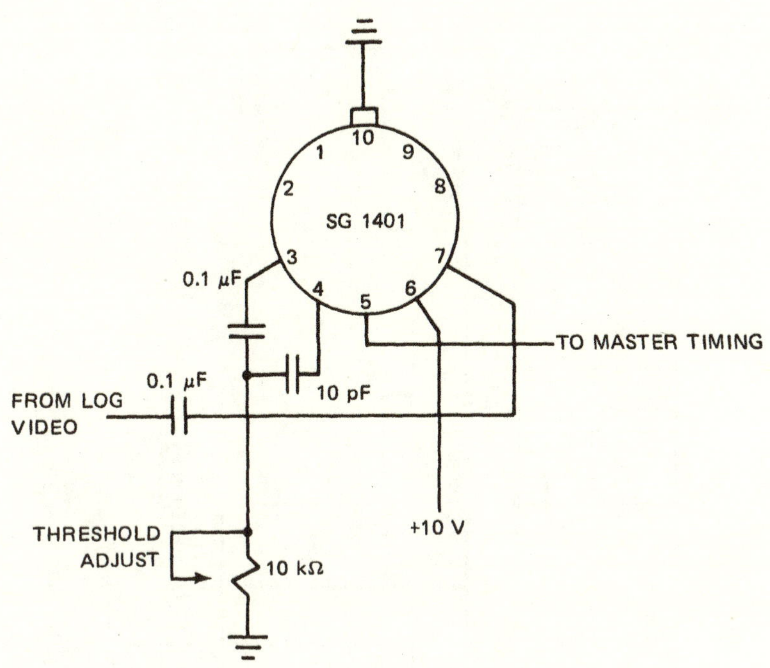

FIGURE 43. AGC Threshold Amplifier.

where the pulse repetition interval (PRI) is

$$PRI = \frac{1}{PRF} \tag{2-34}$$

Thus

$$\tau_{r,SL} = \frac{1.05\ RC}{XA_{\Delta}A_{\epsilon}(LS)T_u(PRF)} \tag{2-35}$$

Now there are N PRF pulses in the loop rise time, $\tau_{r,SL}$

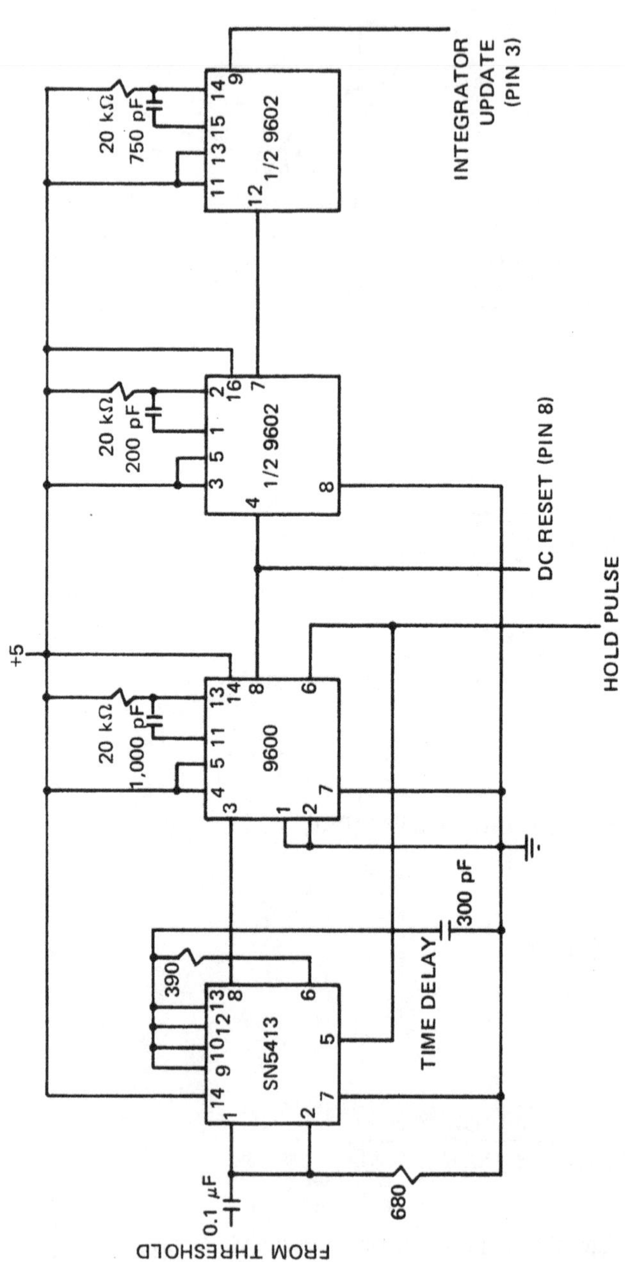

FIGURE 44. AGC Time Delay and Master Timing.

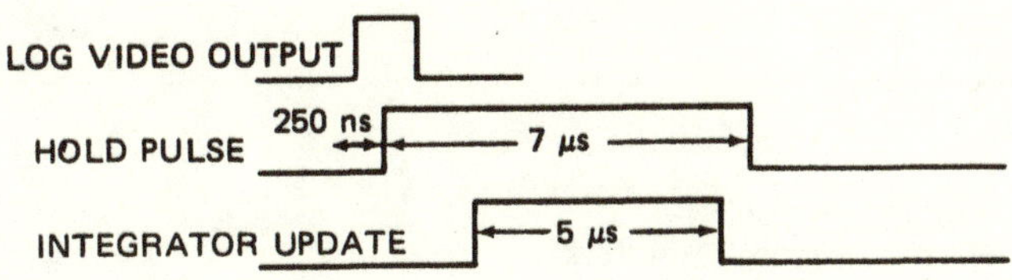

FIGURE 45. Pulse AGC Timing.

$$N = (PRF)\,(\tau_{r,SL}) \tag{2-36}$$

where

N = number of pulses necessary to close loop (10 to 90%)

or

$$\tau_{r,SL} = \frac{N}{PRF} \tag{2-37}$$

Now, substituting Equation 2-37 into 2-35,

$$N = \frac{1.05\ RC}{XA_{\Delta}A_{\epsilon}(LS)T_u} \tag{2-38}$$

A minimum of 65 pulses is required to close the loop. This number will occur for the maximum value of X; thus,

$$RC = \frac{(65)\ (17.5)\ (1)\ (1)\ (0.090)\ (5 \times 10^{-6})}{1.05} \tag{2-39}$$

or

$$RC = 4.88 \times 10^{-4} \tag{2-40}$$

Let $C = 0.1\ \mu F$;

$$R = \frac{4.88 \times 10^{-4}}{0.1 \times 10^{-6}} = 4.88\ k\Omega \tag{2-41}$$

Using a standard value of 4.7 kilohms for R gives results very close to those predicted; $N \cong 70$ pulses minimum ($X \cong 17.5$ dBm/V) and 250 pulses maximum ($X \cong 5$ dBm/V).

Figure 46 illustrates the AGC voltage versus input power. AGC action was adjusted to start (via the AGC delay) at −72 dBm. The static regulation was measured, and the output power varied less than 0.2 dBm over a 50-dBm input dynamic range.

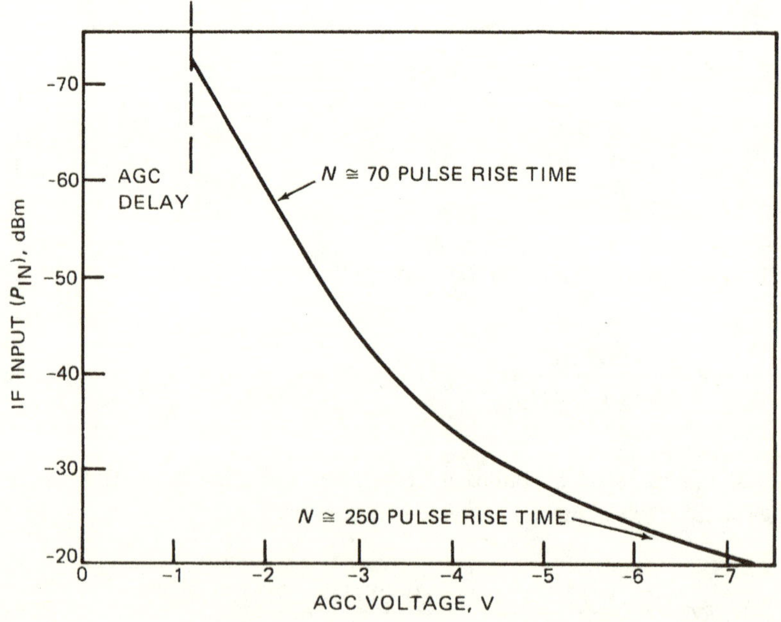

FIGURE 46. AGC Voltage Versus IF Input Power.

This example was presented to illustrate the techniques involved. The nonlinear AGC gain slope does limit the usefulness of this scheme. More consistent results would be obtained using the 30-megahertz IF amplifier illustrated in Figure 22. The reader will have to justify to himself the necessity for employing a logarithmic amplifier for individual design needs.

CONICAL SCAN AGC AND FREQUENCY STABILITY

In presenting the AGC loops we have assumed that a single low-pass filter (or a true integrator) determines the frequency response, and this frequency response is much less than the open-loop frequency response. This is legitimate for many AGC needs, but not all. One notable exception is the AGC loop for conical scan radars.[4,5] Figure 47a illustrates a typical conical scan radar. The angle return is dependent on the modulation envelope of the return pulses (Figure 47b) and phase with respect to a reference voltage (Figure 47c). The resultant azimuth and elevation signals are shown in Figure 47d.

The information necessary for angle tracking is usually obtained by comparing the received video with a reference signal (Figure 47b and c); thus the signal modulation envelope should have no AGC at the scan frequency (if AGC were applied at the scan frequency, no modulation envelope would result due to the inherent input modulation reduction characteristics of the AGC loop). Furthermore, any phase shift of the received modulation envelope must be kept small to avoid cross talk between the azimuth and elevation angle loops. The loop gain at the scan frequency must be much less than unity as will be shown.

The modulation characteristics for an AGC loop have been given as (Equation 1-23)

$$\frac{\Delta e_{IF}(PP)}{e_{IF}(PP)} = \frac{1}{1 + LG} \left(\frac{\Delta e_{\text{in}}(PP)}{e_{\text{in}}(PP)} \right) \tag{2-42}$$

[4] M. I. Skolnik. *Introduction to Radar Systems.* New York, McGraw-Hill Book Co., 1962.

[5] D. K. Barton. *Radar System Analysis.* Englewood Cliffs, N.J., Prentice Hall, Inc., 1964.

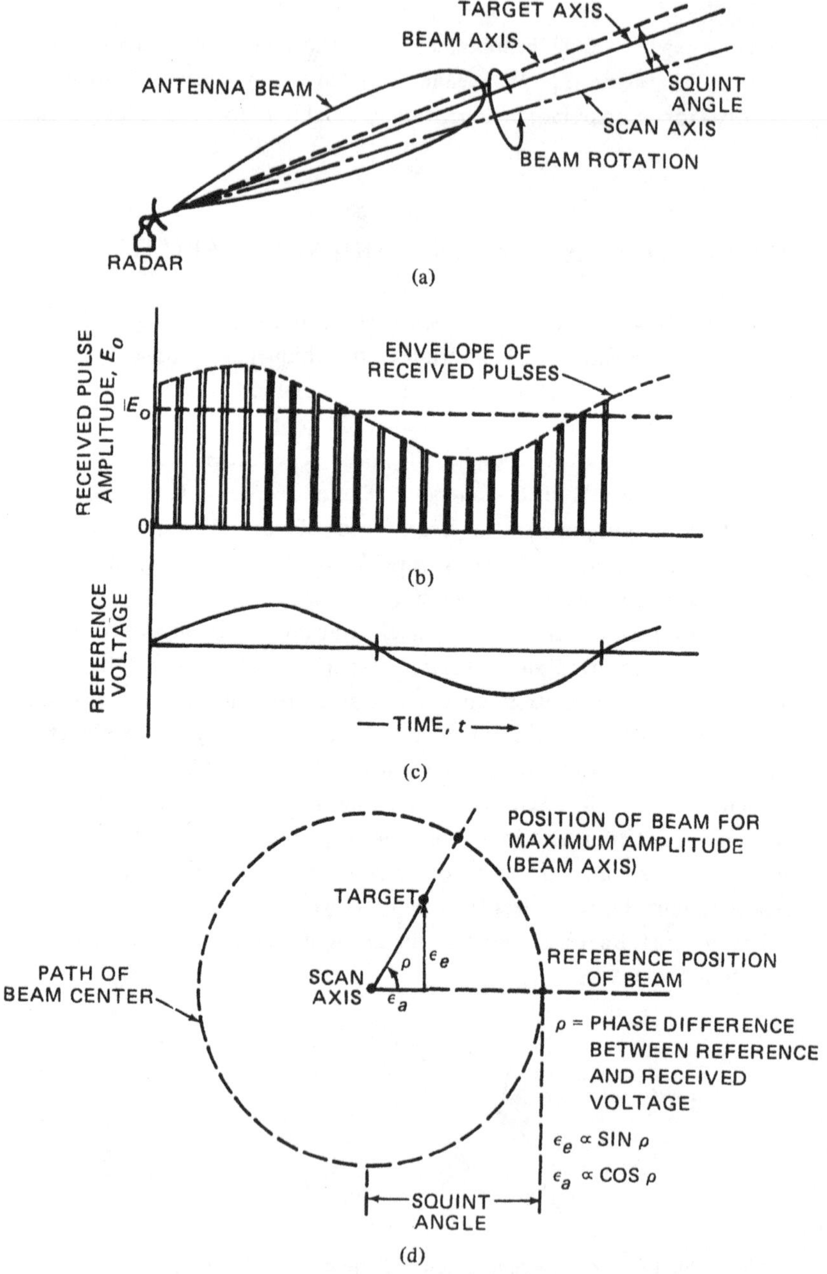

FIGURE 47. Typical Conical Scan Radar. (a) Basic radar, (b) received target envelope, (c) reference voltage, (d) azimuth and elevation signals.

or, defining modulation (M) as

$$M = \Delta e / e \tag{2-43}$$

$$M_{IF} = \frac{M_{\text{in}}}{1 + LG} \tag{2-44}$$

The denominator of Equation 2-42 determines the modulation characteristics of the AGC loop. As mentioned earlier, the phase shift must be small at the conical scan frequency; thus the modulation phase shift (ϕ) may be given as (for small loop gain at the scan frequency)

$$\phi = \tan^{-1} \frac{LG}{1 + LG} \cong \tan^{-1} (LG) \tag{2-45}$$

If, for example the required phase shift for a given conical scan radar must be less than 2 degrees, then the loop gain at the scan frequency must be

$$LG < \tan 2 \text{ deg} \tag{2-46}$$

or

$$LG < 0.0349 \text{ (or } -29.1 \text{ dBV)} \tag{2-47}$$

Thus some method must be obtained to ensure a loop gain of -29.1 dBV at the conical scan frequency. There are several other requirements for conical scan AGC loops, and they may be summarized as follows[6,7]:

[6] D. J. Povejsil, R. S. Raven, and P. Waterman. *Airborne Radar.* New York, D. Van Nostrand Co., Inc., 1961, pp. 419-26.

[7] J. C. G. Field. *The Design of Automatic-Gain-Control Systems for Auto-Tracking Radar Receivers.* The Institution of Electrical Engineers, Monograph No. 258 R, October 1957, pp. 93-108.

1. There must be high gain at low frequencies (static regulation).
2. The AGC bandwidth must be adequate to ensure proper isolation from amplitude fluctuations in the received echo (for modulation frequencies below the scanning frequency).
3. The phase shift must be small at the modulation frequency to minimize cross talk. This determines the loop gain necessary at the scan frequency (Equation 2-45).
4. To ensure adequate loop stability and transient response, a minimum gain margin of 6 dBV must be maintained. Similarly, a phase margin of 45 degrees should be maintained. As the AGC loop is simply that of a classical feedback system, whenever a 180-degree phase shift is obtained for loop gains greater than unity (0 dBV), oscillations will occur.
5. In pulsed systems, it is necessary to provide a minimum gain margin of at least 6 dBV at one-half the pulse repetition frequency to ensure stable operation.[7]

The low-pass-filter loop (Figure 2a) is the most practical one to use in conical scan AGC. A simple example will now be given to illustrate the techniques involved.[7]

A conical scan AGC is needed that has the following parameters:

$P_{in,min} = -90$ dBm $\quad P_{in,max} = +10$ dBm

ΔP_o(dBm) $< \pm 1$ dBm $\quad$ linear detector

Conical scan frequency $\cong 30$ Hz $\quad$ PRF = 2 kHz

$\phi < 2$ degrees $\quad \Delta AGC = 5$ volts

$e_N = 1$ volt

The total input dynamic range is

$$\Delta P_{in}(\text{dBm}) = -90 + 10 = 100 \text{ dBm} \tag{2-48}$$

Using Figure 20,

$$A_\Delta A_\epsilon e_N = \frac{\Delta AGC}{10^{\frac{n\Delta P_o(\mathrm{dBm})}{20}} - 1} \tag{2-49}$$

since

$$n = 1 \text{ (linear detector)} \tag{2-50}$$

$$A_\Delta A_\epsilon e_N = \frac{5}{10^{\frac{1}{20}} - 1} \tag{2-51}$$

The loop gain may now be given as (Figure 20)

$$LG = 0.12\ nXA_\Delta A_\epsilon e_N \tag{2-52}$$

where

$$X = \mathrm{dBm/V} = 100/5 = 20 \tag{2-53}$$

Thus

$$LG = (0.12)\ (1)\ (20)\ 41 = 98.4 \text{ (or 39.9 dBV)} \tag{2-54}$$

Equation 2-45 relates the loop gain necessary to give the wanted maximum phase shift at the conical scan frequency,

$$LG < \tan \phi_M \tag{2-55}$$

or

$$LG < -29.1 \text{ dBV} \tag{2-56}$$

Thus the loop gain must be 39.9 dBV at low frequencies and decrease to -29.1 dBV (69 dBV change) at the conical scan frequency, 30 hertz. Also, any phase shift must be less than 180 degrees for any loop gain greater than 0 dBV, and the loop gain must be less than -6 dBV at one-half the PRF (or loop gain must be less than -6 dBV at 1000 hertz). The easiest solution is to use a twin-T filter in series with a suitable low-pass filter. It should be pointed out that the variable gain slope, X, is usually not linear; thus the loop gain will vary as X varies, and the designer must ensure that the loop remains stable. Field[7] has an excellent discussion on filter techniques necessary for conical scan radar AGC stability.

CRYSTAL VIDEO AGC

Radar receivers are sometimes required to have a very large instantaneous bandwidth. One simple method of achieving this is with a crystal video receiver as shown in Figure 48.

The basic operation and analysis of the crystal video AGC loop are exactly the same as for the loops previously described, but without the added complexity of the detector (the detector is outside the AGC loop), derivations of the pertinent design equations are exactly the same as outlined in Appendices C, D, and E, and are summarized in Figure 48 and 50.

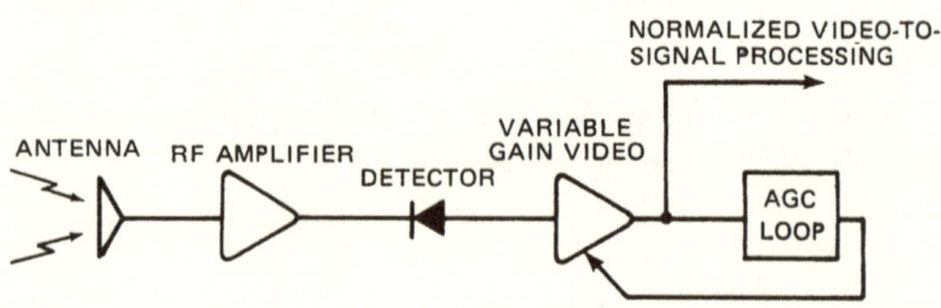

FIGURE 48. Basic Crystal Video Receiver.

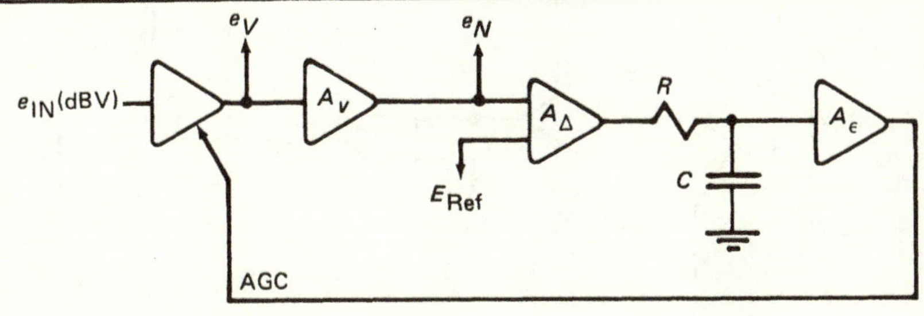

Static Regulation

$$\Delta e_N(\text{dBV}) = 20 \log \left(\frac{\Delta AGC}{A_\Delta A_\epsilon e_N} + 1 \right)$$

$$A_\Delta A_\epsilon e_N = \frac{\Delta AGC}{10^{\frac{\Delta e_N(\text{dBm})}{20}} - 1}$$

Dynamic Regulation

$$LG = 0.115\, X A_\Delta A_\epsilon e_N$$

$$A_\Delta A_\epsilon e_N = \frac{LG}{0.115\, X}$$

Loop Rise Time (10 to 90%)

$$\tau_r = \frac{18.3\, RC}{X A_\Delta A_\epsilon D e_N} \qquad RC = 0.055\, \tau_r X A_\Delta A_\epsilon D e_N$$

$$f_{3\text{dBV}}(LG) = \frac{0.159\, D}{RC} \qquad (LG)\,(\tau_r) = \frac{2.1\, RC}{D}$$

X = variable gain slope (dBV/V) $\qquad$ D = LPF update duty cycle

FIGURE 49. Low-Pass Filter, Crystal Video AGC Loop Design Equations.

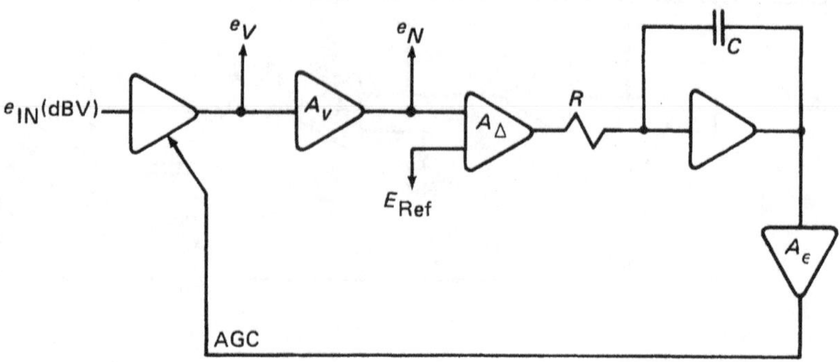

Static Regulation

$$\Delta e_N(\text{dBV}) \cong 0$$

Dynamic Regulation

$$LG = \frac{0.115\, XA_\Delta A_\epsilon e_N}{fRC} \qquad A_\Delta A_\epsilon e_N = \frac{LG\, f\, RC}{0.115\, X}$$

Loop Rise Time (10 to 90%)

$$\tau_r = \frac{18.3\, RC}{XA_\Delta A_\epsilon De_N} \qquad RC = 0.055\ \tau_r XA_\Delta A_\epsilon De_N$$

X = variable gain slope (dBV/V) $\qquad$ D = integrator update duty cycle

FIGURE 50. Integrator, Crystal Video AGC Loop Design Equations.

Appendix A

POWER-VOLTAGE RELATIONSHIPS FOR A 50-OHM SYSTEM

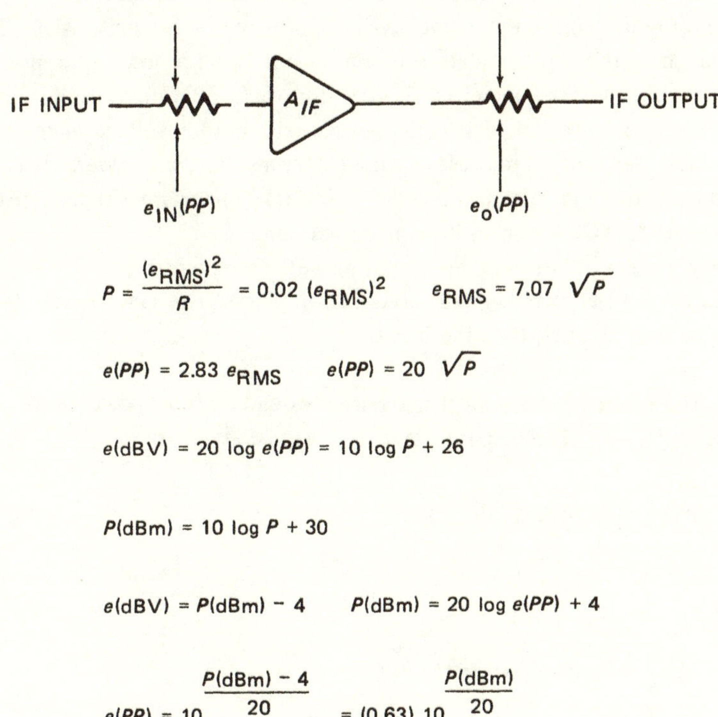

FIGURE A-1. Power-Voltage Relationships for a 50-Ohm System.

Appendix B

BASIC DETECTOR CHARACTERISTICS

The crystal diode detector converts the IF waveform into a DC waveform for continuous wave AGC, or into pulses for pulse AGC. The basic characteristics for detectors are *square law* for low input powers (the output voltage increases 2 dBV for each dBM increase in input power), and *linear* for high input powers (the output voltage increases 1 dBV for each dBm increase in input power). Watson[2] covers detector characteristics in detail; however, several important characteristics pertinent to AGC design will be presented here.

Figure B-1 illustrates the input-output characteristics for a typical Schottky barrier detector (HP 5082-2800). The square law characteristics will be covered first, then the linear.

The square law detector characteristics extend for input power of -8 dBm and lower. The output voltage, e_D, may be given as

$$e_D = K_{SL}\ [e_{PD}(PP)]^2 \qquad \text{(B-1)}$$

where

K_{SL} = diode constant

The diode constant is

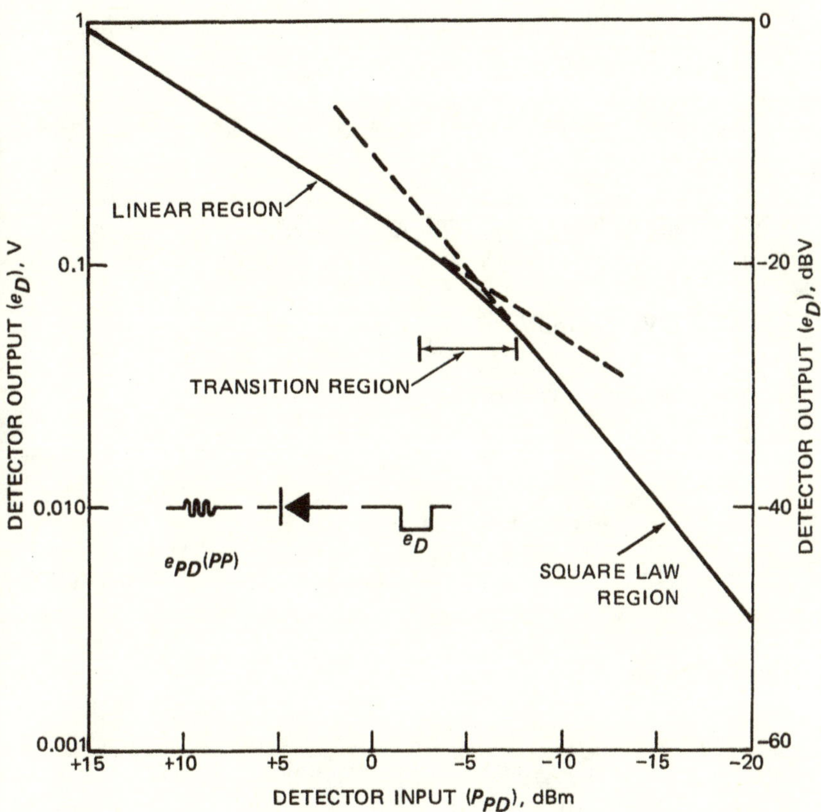

FIGURE B-1. Detector Characteristics.

$$K_{SL} = \frac{e_D}{[e_{PD}(PP)]^2} \tag{B-2}$$

A curve relating e_D to $e_{PD}(PP)$ is seldom, if ever, given. The value of K_{SL} may be found, provided a curve similar to that in Figure B-1 is given, as follows:

Using Equation B-1,

$$20 \log e_{D,SL} = e_D(\text{dBV}) = K_{SL}(\text{dBV}) + 40 \log e_{PD}(PP) \tag{B-3}$$

or

$$K_{SL}(\text{dBV}) = e_D(\text{dBV}) = -40 \log e_{PD}(PP) \tag{B-4}$$

Substituting $P(\text{dBm})$ for $e_{PD}(PP)$ (Appendix A),

$$K_{SL}(\text{dBV}) = e_D(\text{dBV}) - 2\, P_{PD}(\text{dBm}) + 8 \tag{B-5}$$

or

$$K_{SL} = 10^{\frac{e_D(\text{dBV}) - 2\, P_{PD}(\text{dBm}) + 8}{20}} \tag{B-6}$$

Simplifying,

$$K_{SL} = 2.51 \left[10^{\frac{e_D(\text{dBV})}{20}}\right] \left[10^{\frac{-P_{PD}(\text{dBm})}{10}}\right] \tag{B-7}$$

and simplifying further,

$$K_{SL} = 2.51\, e_D\, 10^{\frac{-P_{PD}(\text{dBm})}{10}} \tag{B-8}$$

Referring to Figure B-1, the square-law diode constant is ($P_{PD}(\text{dBm}) = -20$, $e_D = 3.5$ mV)

$$K_{SL} = (2.51)\,(3.5 \times 10^{-3})\, 10^{\frac{-(-20)}{10}} = 0.879 \tag{B-9}$$

Knowing K_{SL}, e_D for any P_{PD}(dBm) may be found by solving Equation B-8 for e_D:

$$e_D = 0.4\ K_{SL}\ 10^{\frac{P_{PD}(\text{dBm})}{10}} \qquad \text{(B-10)}$$

or, if e_D(dBV) is wanted,

$$e_D(\text{dBV}) = K_{SL}(\text{dBV}) + 2\ P_{PD}(\text{dBm}) - 8 \qquad \text{(B-11)}$$

The dynamic detector gain, $A_{D,SL}$(V/V) is important in finding the loop gain. The dynamic detector gain may be found for a given normalized input, $e_{PD}(PP)$, by differentiating Equation B-2:

$$A_{D,SL}(\text{V/V}) = \frac{d\ e_D}{d\ e_{PD}(PP)} \quad (e_{PD}(PP) = \text{constant}) \qquad \text{(B-12)}$$

or

$$A_{D,SL}(\text{V/V}) = 2\ K_{SL}\, e_{PD}(PP) \qquad \text{(B-13)}$$

Referring again to Appendix A for $e(PP)$,

$$A_{D,SL}(\text{V/V}) = 2\ K_{SL}\ 10^{\frac{P_{PD}-4}{20}} \qquad \text{(B-14)}$$

or

$$A_{D,SL}(\text{V/V}) = 1.26\ K_{SL}\ 10^{\frac{P_{PD}(\text{dBm})}{20}} \qquad \text{(B-15)}$$

The dynamic detector gain in V/dBm, $A_{D,SL}$(V/dBm), is needed when solving for the loop rise time, τ_r. This gain may be found by differentiating Equation B-10. The solution to this differentiation is

$$A_{D,SL}(\text{V/dBm}) = 0.092\ K_{SL}\ 10^{\frac{P_{PD}(\text{dBm})}{10}} \qquad \text{(B-16)}$$

The voltage gains for the diode illustrated in Figure B-1 are, assuming P_{PD}(dBm) = −30 dBm,

$$A_{D,SL}(\text{V/V}) = (1.26)\ (0.879)\ 10^{\frac{-30}{20}} = 0.035\ \text{V/V} \qquad \text{(B-17)}$$

$$A_{D,SL}(\text{V/dBm}) = (0.092)\ (0.879)\ 10^{\frac{-30}{10}} = 8.09 \times 10^{-5}\ \text{V/dBm} \qquad \text{(B-18)}$$

Figure B-2 summarizes the square law detector characteristics.

Linear detector characteristics extend for input powers larger than -2 dBm for the HP 5082-2800 detector (Figure B-1). The output-voltage may be given as

$$e_D = K_{Lin}\ e_{PD}(PP) \qquad \text{(B-19)}$$

where

K_{Lin} = diode constant

Since, using Appendix A,

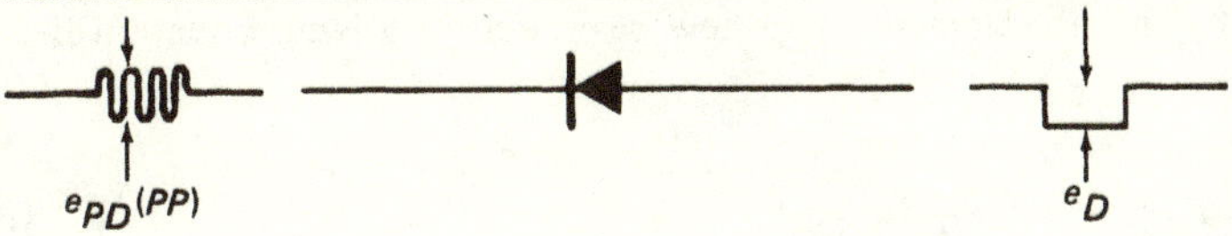

$e_{PD}(PP)$ e_D

$$e_D = K_{SL}\left[e_{PD}(PP)\right]^2 = 0.4\,K_{SL}\,10^{\frac{P_{PD}(\text{dBm})}{10}}$$

$$P_{PD}(\text{dBm}) = 10\log e_D - 10\log K_{SL} + 4 = \frac{e_D(\text{dBV})}{2} - K_{SL}(\text{dBV}) + 4$$

$$P_{PD}(\text{dBm}) = 10\log\left[\frac{0.25\,e_D}{K_{SL}}\right]$$

$$K_{SL} = \frac{e_D}{\left[e_{PD}(PP)\right]^2} = 2.51\,e_D\,10^{\frac{-P_{PD}(\text{dBm})}{10}}$$

$$A_{D,SL}(\text{V/V}) = 2\,K_{SL}\,e_{PD}(PP) \qquad \text{V/V}$$

$$A_{D,SL}(\text{V/V}) = 1.26\,K_{SL}\,10^{\frac{P_{PD}(\text{dBm})}{20}} \qquad \text{V/V}$$

$$A_{D,SL}(\text{V/dBM}) = 0.092\,K_{SL}\,10^{\frac{P_{PD}(\text{dBm})}{10}} \qquad \text{V/dBm}$$

$$A_{D,SL}(\text{V/dBm}) = 0.23\,e_D \qquad \text{V/dBm}$$

FIGURE B-2. Square Law Detector Characteristics.

$$e_{PD}(PP) = (0.63)\,10^{\frac{P_{PD}(\text{dBm})}{20}} \tag{B-20}$$

Equation B-19 becomes

$$e_D = 0.63\,K_{Lin}\,10^{\frac{P_{PD}(\text{dBm})}{20}} \tag{B-21}$$

K_{Lin} for the detector may now be found by solving Equation B-21:

$$K_{Lin} = 1.6\, e_D\, 10^{\frac{-P_{PD}(\text{dBm})}{20}} \tag{B-22}$$

Referring to Figure B-1, the linear detector diode constant for $P_{PD}(\text{dBm}) = +10$, $e_D = 0.52$, is

$$K_{Lin} = (1.6)\,(0.52)\,10^{\frac{-10}{20}} = 0.26 \tag{B-23}$$

The dynamic detector gain, $A_{D,Lin}(\text{V/V})$, is

$$A_{D,Lin}(\text{V/V}) = \frac{d\,e_D}{d\,e_{PD}(PP)} \qquad (e_{PD}(PP) = \text{constant}) \tag{B-24}$$

or

$$A_{D,Lin}(\text{V/V}) = K_{Lin} \tag{B-25}$$

Thus for the diode under discussion,

$$A_{D,Lin}(\text{V/V}) = 0.26 \text{ V/V} \tag{B-26}$$

The dynamic detector gain in V/dBm, $A_{D,Lin}(\text{V/dBm})$ can be found by differentiating Equation B-21,

$$A_{D,Lin}(\text{V/dBm}) = 0.073\, K_{Lin}\, 10^{\frac{P_{PD}(\text{dBm})}{20}} \tag{B-27}$$

and for our detector is (P_{PD}(dBm) = +10)

$$A_{D,Lin}(\text{V/dBm}) = (0.073)\,(0.26)\,10^{\frac{10}{20}} = 0.0623 \text{ V/dBm} \qquad \text{(B-28)}$$

Figure B-3 summarizes linear detector characteristics.

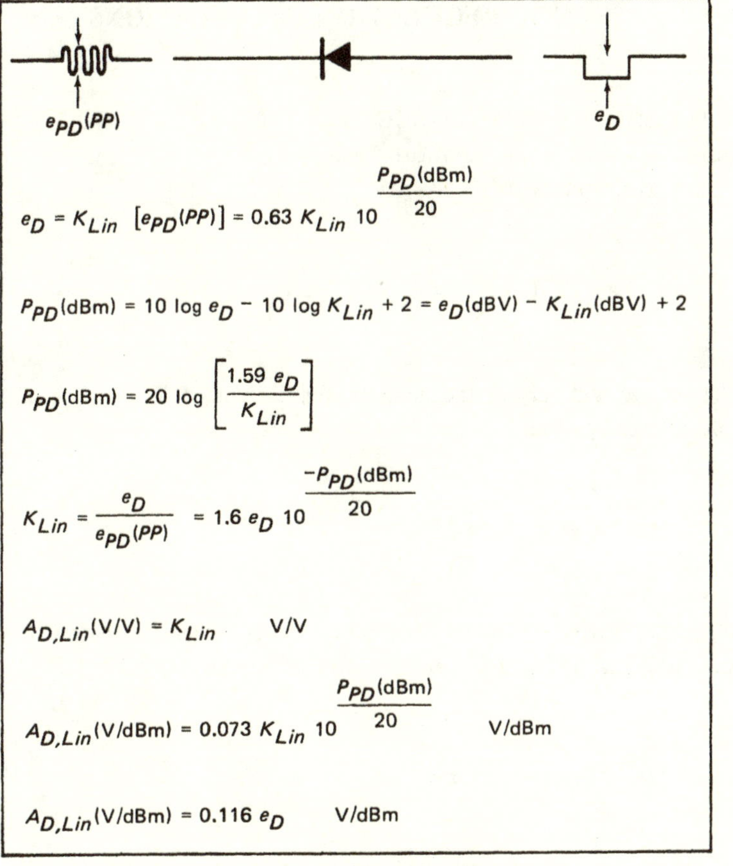

FIGURE B-3. Linear Detector Characteristics.

Appendix C

STATIC REGULATION CALCULATIONS

The static regulation of AGC loops will now be given. Figure 1a illustrates the basic loop configuration.

The AGC voltage is

$$AGC = A_{\Delta} A_{\epsilon} (E_{\mathrm{Ref}} - A_{\nu} e_D) \tag{C-1}$$

The voltage will vary as the input power is changed. The change in AGC voltage may be given as

$$\Delta AGC = A_{\Delta} A_{\epsilon} A_{\nu} \Delta e_D \tag{C-2}$$

where Δe_D is the maximum permissible change in detector output for the wanted change in input power. Δe_D may be given as

$$\Delta e_D = e_{D,\max} - e_{D,\min} \tag{C-3}$$

e_D is related to detector input power, for a square law detector, by (Appendix B)

$$e_D = 0.4\, K_{SL}\ 10^{\frac{P_{PD}(\mathrm{dBm})}{10}} \qquad \text{(C-4)}$$

Substituting Equation C-4 into C-3,

$$\Delta e_D = 0.4\, K_{SL} \left[10^{\frac{P_{PD,\max}(\mathrm{dBm})}{10}} - 10^{P_{PD,\min}(\mathrm{dBm})} \right] \qquad \text{(C-5)}$$

This equation may be rewritten as

$$\Delta e_D = 0.4\, K_{SL}\ 10^{\frac{P_{PD,\min}(\mathrm{dBm})}{10}} \left[\frac{10^{\frac{P_{PD,\max}(\mathrm{dBm})}{10}}}{10^{P_{PD,\min}(\mathrm{dBm})}} - 1 \right] \qquad \text{(C-6)}$$

or

$$\Delta e_D = e_{D,\min} \left[10^{\frac{\Delta P_{PD}(\mathrm{dBm})}{10}} - 1 \right] \qquad \text{(C-7)}$$

where

$$\Delta P_{PD,SL}(\mathrm{dBm}) = P_{PD,\max}(\mathrm{dBm}) - P_{PD,\min}(\mathrm{dBm})$$

Substituting Equation C-7 into C-2 yields

$$\Delta AGC = A_\nu A_\Delta A_\epsilon e_{D,\min} \left[10^{\frac{\Delta P_{PD,SL}(\mathrm{dBm})}{10}} - 1 \right] \qquad \text{(C-8)}$$

and solving Equation C-8 for ΔP_{PD}(dBm) gives

$$\Delta P_{PD,SL}(\text{dBm}) = 10 \log \left(\frac{\Delta AGC}{A_v A_\Delta A_\epsilon e_{D,\min}} + 1 \right) \tag{C-9}$$

Since the variable gain IF and predetector IF amplifiers are linear,

$$\Delta P_{PD,SL}(\text{dBm}) = \Delta P_{IF}(\text{dBm}) \tag{C-10}$$

and

$$\Delta P_{IF,SL}(\text{dBm}) = 10 \log \left(\frac{\Delta AGC}{A_v A_\Delta A_\epsilon e_{D,\min}} + 1 \right) \tag{C-11}$$

The average value for $e_{D,N}$ is

$$e_{D,N} = \frac{e_{D,\max} + e_{D,\min}}{2} \tag{C-12}$$

If $e_{D,N}$ is adjusted in the mid-AGC range, $\dfrac{\Delta AGC}{2}$,*

$$\Delta P_{IF,SL}(\text{dBm}) = 10 \log \left(\frac{\Delta AGC}{A_v A_\Delta A_\epsilon e_{D,N}} + 1 \right) \tag{C-13}$$

The same procedure can be used to find the static regulation for linear detectors:

* If the AGC voltage does not start at zero volts, ΔAGC/2 must be added to the voltage at which the AGC action does start.

$$\Delta P_{IF,Lin}(\text{dBm}) = 20 \log \left(\frac{\Delta AGC}{A_v A_\Delta A_\epsilon e_{D,N}} + 1 \right) \tag{C-14}$$

Noting that

$$A_v e_{D,N} = e_N \tag{C-15}$$

Equations C-13 and C-14 become

$$\Delta P_{IF,SL}(\text{dBm}) = 10 \log \left(\frac{\Delta AGC}{A_\Delta A_\epsilon e_N} + 1 \right) \tag{C-16}$$

and

$$\Delta P_{IF,Lin}(\text{dBm}) = 20 \log \left(\frac{\Delta AGC}{A_\Delta A_\epsilon e_N} + 1 \right) \tag{C-17}$$

The IF output may be divided (power split) to drive two or more circuits. Provided all circuitry is linear, the change in power-split output, ΔP_o(dBm), will be the same as ΔP_{IF}(dBm).

Appendix D

AGC GAIN CALCULATION

The loop gain of the AGC loop is dependent on the AGC gain of the variable gain IF amplifier. The AGC gain is defined as the change in IF output voltage for a given change in AGC voltage, or

$$A_{AGC} = \frac{\Delta e_{IF}(PP)}{\Delta AGC} \qquad (P_{\text{in}}(\text{dBm}) = \text{constant}) \tag{D-1}$$

where

$\Delta e_{IF}(PP) = e_{IF,\max}(PP) - e_{IF,\min}(PP)$

$\Delta AGC = AGC_{\max} - AGC_{\min}$

$AGC_{\min}$ = AGC voltage for minimum gain

$AGC_{\max}$ = AGC voltage for maximum gain

The IF output power is

$$P_{IF}(\text{dBm}) = X(AGC) \tag{D-2}$$

The IF output voltage in terms of output power is (Appendix A)

$$e_{IF}(PP) = (0.63)\ 10^{\frac{P_{IF}(\text{dBm})}{20}} \tag{D-3}$$

or, substituting Equation D-2,

$$e_{IF}(PP) = (0.63)\ 10^{\frac{X(AGC)}{20}} \tag{D-4}$$

The change in amplifier output voltage now may be written as

$$\Delta e_{IF}(PP) = 0.63 \left[10^{\frac{X(AGC)_{\max}}{20}} - 10^{\frac{X(AGC)_{\min}}{20}} \right] \tag{D-5}$$

or

$$\Delta e_{IF}(PP) = 0.63 \left(10^{\frac{X(AGC)_{\min}}{20}} \right) \left[\frac{10^{\frac{X(AGC)_{\max}}{20}}}{10^{\frac{X(AGC)_{\min}}{20}}} - 1 \right] \tag{D-6}$$

or, using Equation D-4 and simplifying,

$$\Delta e_{IF}(PP) = e_{IF,\min}(PP) \left(10^{\frac{X \Delta AGC}{20}} - 1 \right) \tag{D-7}$$

The parenthetical term in Equation D-7 is of the general form a^Y. Expanding this exponential function,

$$a^Y = 1 + Y \ln (a) + \left[\frac{Y \ln (a)}{2!} \right]^2 + \left[\frac{Y \ln (a)}{3!} \right]^3 + \ldots \tag{D-8}$$

Provided Y is small ($Y < 0.5$), Equation D-8 simplifies to

$$a^Y \cong 1 + Y \ln(a) \qquad \text{(D-9)}$$

Equation D-7 now becomes

$$\Delta e_{IF}(PP) = e_{IF,\min}(PP) \left[\frac{X\Delta AGC}{20} \quad (\ln 10)\right] \qquad \text{(D-10)}$$

or

$$\Delta e_{IF}(PP) = 0.115\ X\Delta AGC\ e_{IF,\min}(PP) \qquad \text{(D-11)}$$

Now Equation D-11 may be solved to find the AGC gain,

$$A_{AGC} = 0.115 X e_{IF,\min}(PP) \qquad \text{(D-12)}$$

and this equation is valid, to within 10%, for

$$\frac{X\Delta AGC}{20} \quad < 0.5 \qquad \text{(D-13)}$$

The $e_{IF,\min}(PP)$ term may be replaced by $e_{IF}(PP)$; thus,

$$A_{AGC} = 0.115 X e_{IF}(PP) \qquad \text{(D-14)}$$

Appendix E

LOOP RISE TIME CALCULATIONS

The loop rise time for the integrator AGC loop (Figure 2b) will now be found. The general method will be to solve for the AGC voltage in terms of input power. This equation can then be differentiated to find the change in AGC voltage with change in input power.

The AGC error voltage, e_ϵ, is zero under static conditions; however, if the input power is increased, the error voltage will increase and fall back toward zero as the loop nulls. This error voltage may be given as

$$e_\epsilon = A_\nu A_\Delta e_D - A_\Delta E_{\text{Ref}} \tag{E-1}$$

The value for e_D may be given as $e_{D,N}$, the normalized or static value, plus Δe_D, the change in e_D due to a change in input power. The static value for e_D may be given, for a square law detector, as (Appendix B)

$$e_{D,N} = 0.4\, K_{SL}\, 10^{\frac{P_{PD,N}(\text{dBm})}{10}} \tag{E-2}$$

The value for Δe_D is

$$\Delta e_D = [A_{D,SL}(\text{V/dBm})]\,[\Delta P_{PD}(\text{dBm})] \tag{E-3}$$

or, from Appendix B,

$$\Delta e_D = 0.23\ e_{D,N}\ \Delta P_{PD}(\text{dBm}) \tag{E-4}$$

where

$e_{D,N}$ = normalized detector output voltage

The instantaneous detector output may now be given as

$$e_D = e_{D,N} + 0.23\ e_{D,N}\ [\Delta P_{PD}(\text{dBm})] \tag{E-5}$$

The value for $\Delta P_{PD}(\text{dBm})$ may be given as

$$\Delta P_{PD}(\text{dBm}) = P_{PD}(\text{dBm}) - P_{PD,N}(\text{dBm}) \tag{E-6}$$

since

$$P_{PD}(\text{dBm}) = P_{\text{in}}(\text{dBm}) + A_{IF}(\text{dBm}) + A_{PD}(\text{dBm}) \tag{E-7}$$

and (Figure 3)

$$A_{IF}(\text{dBm}) = A_o(\text{dBm}) - X(AGC) \tag{E-8}$$

$$P_{PD}(\text{dBm}) = P_{\text{in}}(\text{dBm}) + A_o(\text{dBm}) - X(AGC) + A_{PD}(\text{dBm}) \tag{E-9}$$

Equation E-5 now becomes

$$e_D = e_{D,N} \Big\{ 1 + 0.23\, [P_{\text{in}}(\text{dBm}) + A_0(\text{dBm}) - X(AGC) + A_{PD}(\text{dBm}) - P_{PD,N}(\text{dBm})] \Big\} \tag{E-10}$$

The error voltage may now be written as

$$e_\epsilon = A_v A_\Delta \left(e_{D,N} \Big\{ 1 + 0.23\, [P_{\text{in}}(\text{dBm}) + A_0(\text{dBm}) - X(AGC) + A_{PD}(\text{dBm}) - P_{PD,N}(\text{dBm})] \Big\} - \frac{E_{\text{Ref}}}{A_v} \right) \tag{E-11}$$

The integrator output is

$$AGC(S) = \frac{e_\epsilon(S) A_\epsilon}{RC\,S} \tag{E-12}$$

Substituting Equation E-12 into E-11 and solving for $AGC(S)$,

$$AGC(S) = A_v A_\Delta A_\epsilon \left(e_{D,N} \Big\{ 1 + 0.23\, [P_{\text{in}}(\text{dBm}) + A_0(\text{dBm}) + A_{PD}(\text{dBm}) - X AGC(S) - P_{PD,N}(\text{dBm})] - \frac{E_{\text{Ref}}}{A_v} \Big\} \right) \Big/ (RC\,S + 0.23\, X A_v A_\Delta A_\epsilon e_{D,N}) \tag{E-13}$$

To find the change in AGC voltage for a given change in input power, Equation E-13 is differentiated with respect to $P_{\text{in}}(\text{dBm})$, giving

$$\frac{\Delta AGC(S)}{\Delta P_{\text{in}}(\text{dBm})(S)} = \frac{0.23\, A_v A_\Delta A_\epsilon e_{D,N}}{RC\,S + 0.23\, X A_v A_\Delta A_\epsilon e_{D,N}} \tag{E-14}$$

or

$$\Delta AGC(S) = \frac{0.23\, A_\nu A_\Delta A_\epsilon e_{D,N}}{RC\, S + 0.23\, XA_\nu A_\Delta A_\epsilon e_{D,N}} \quad \Delta P_{\text{in}}(\text{dBm})\,(S) \qquad \text{(E-15)}$$

Assuming a step change in input power,

$$\Delta AGC(S) = \frac{0.23\, A_\nu A_\Delta A_\epsilon e_{D,N}}{RC\, S + 0.23\, XA_\nu A_\Delta A_\epsilon e_{D,N}} \left[\frac{\Delta P_{\text{in}}(\text{dBm})\,(S)}{S}\right] \qquad \text{(E-16)}$$

The inverse Laplace transform of Equation E-16 is

$$AGC(t) = \frac{\Delta P_{\text{in}}(\text{dBm})}{X} \left[1 - \exp - \left(\frac{t/RC}{0.23\, XA_\nu A_\Delta A_\epsilon e_{D,N}}\right)\right] \qquad \text{(E-17)}$$

The time constant for Equation E-17 is

$$T = \frac{RC}{0.23\, XA_\nu A_\Delta A_\epsilon e_{D,N}} \qquad \text{(E-18)}$$

The 10 to 90% rise time, τ_r, may be approximated as

$$\tau_r \cong 2.1\, T \qquad \text{(E-19)}$$

Thus

$$\tau_{r,SL} = \frac{9.13\, RC}{XA_\nu A_\Delta A_\epsilon e_{D,N}} \qquad \text{(E-20)}$$

or, since

$$A_v e_{D,N} = e_N \tag{E-21}$$

$$\tau_{r,SL} = \frac{9.13\ RC}{XA_\Delta A_\epsilon e_N} \tag{E-22}$$

Using the same procedure, the loop rise time for the linear detector is

$$\tau_{r,Lin} = \frac{18.3\ RC}{XA_\Delta A_\epsilon e_N} \tag{E-23}$$

The loop rise time for the low-pass-filter AGC loop is similar to that for the integrator AGC loop. One main difference, however, is that the error voltage is zero for the integrator AGC loop under static conditions, and is finite for the low-pass-filter loop. The error voltage is small for a well-designed low-pass-filter AGC loop and may usually be neglected. The AGC voltage, for a square law detector, may be given as

$$\Delta AGC(S) = \frac{0.23\ A_v A_\Delta A_\epsilon e_{D,N}}{1 + RCS + 0.23\ XA_v A_\Delta A_\epsilon e_{D,N}} \quad \Delta P_{\text{in}}(\text{dBm})\ (S) \tag{E-24}$$

The AGC voltage for a step change in input power is

$$\Delta AGC(S) = \frac{0.23\ A_v A_\Delta A_\epsilon e_{D,N}}{RCS + 0.23\ XA_v A_\Delta A_\epsilon e_{D,N} + 1} \left[\frac{\Delta P_{\text{in}}(\text{dBm})}{S}\right] \tag{E-25}$$

If

$$0.23\ XA_v A_\Delta A_\epsilon e_{D,N} >> 1 \tag{E-26}$$

$$\Delta AGC(S) = \frac{0.23\, A_\nu A_\Delta A_\epsilon e_{D,N}}{RCS + 0.23\, XA_\nu A_\Delta A_\epsilon e_{D,N}} \tag{E-27}$$

which is exactly the same equation as the integrator AGC loop, Equation E-16. Thus, provided the conditions of Equation E-20 are met, the low-pass-filter AGC loop has the same loop rise time as the integrator AGC loop, Equations E-22 and E-23.

Appendix F

BASICS OF LOGARITHMIC AMPLIFICATION

A logarithmic amplifier has an input-output relationship given by

$$e_{out} = K_1 \log K_2 \, e_{in} \qquad \text{(F-1)}$$

where

K_1 = slope

K_2 = logarithmic offset

There are several ways to consider the relationship expressed in Equation F-1. Figure F-1a illustrates Equation F-1 on linear scales (for two values of K_1 and K_2). (For simplicity $0 < e_{in} < 10$, $K_1 = 0.5$ and 1, and $K_2 = 1$ and 2.) As will be noted, the output is compressed for increasing inputs. Figure F-2b also illustrates Equation F-1, but with the abscissa on a logarithmic scale ($\log e_{in}$). As is to be expected, the input-output relationship is a straight line. Rewriting Equation F-1,

$$e_{out} = K_1 \log e_{in} + K_1 \log K_2 \qquad \text{(F-2)}$$

one can easily see that increasing K_2 will increase the output for any input by a constant term ($K_1 \log K_2$) and will have no effect on the

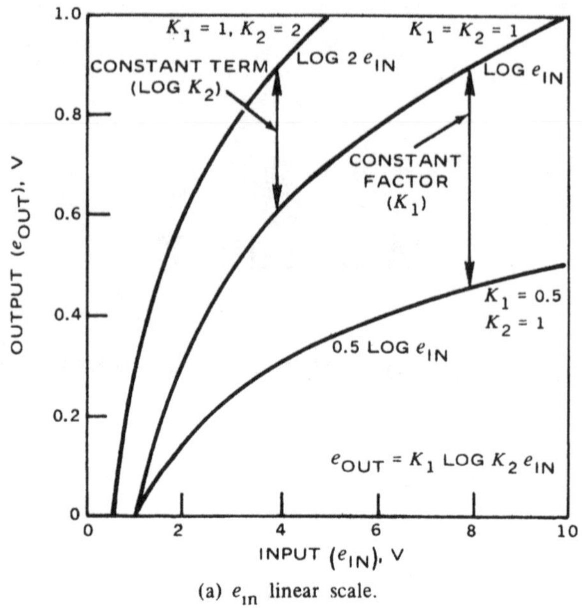

(a) e_{in} linear scale.

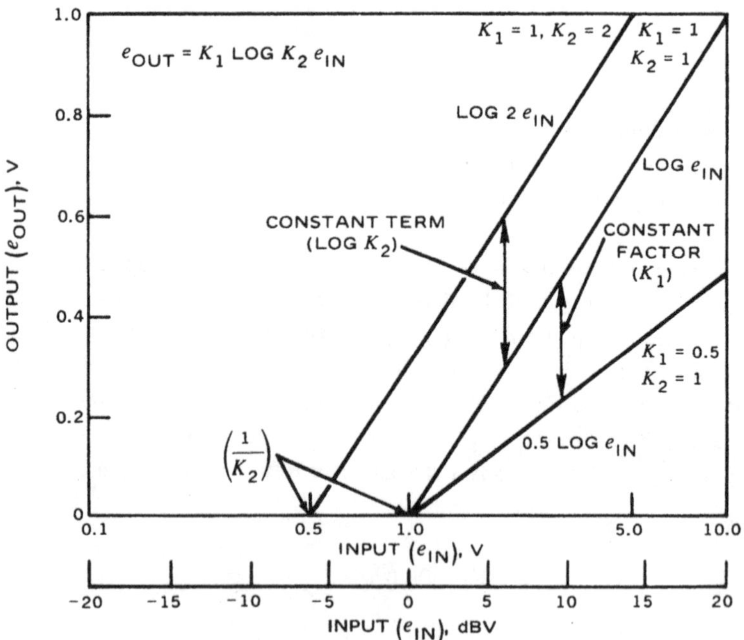

(b) e_{in} logarithmic and decibel scale

FIGURE F-1. Characteristics of Equation F-1.

slope, K_1. Increasing K_1 will increase the slope and increase the output by a constant factor (K_1).

When dealing with logarithmic amplifiers, the output logarithmic slope (*LS*) is usually specified in V/dBV. This slope may be defined with the aid of Figure F-1b (dBV scale) as

$$e_{out,1} = K_1 \log K_2 \, e_{in,1} \tag{F-3a}$$

$$e_{out,2} = K_1 \log K_2 \, e_{in,2} \tag{F-3b}$$

Defining Δe_{out} as

$$\Delta e_{out} = e_{out,2} - e_{out,1} \tag{F-4a}$$

or using Equations F-3a and F-3b,

$$\Delta e_{out} = K_1 \left(\log \frac{e_{in,2}}{e_{in,1}} \right) \tag{F-4b}$$

letting

$$\frac{e_{in,2}}{e_{in,1}} = 10 \text{ (or 1 decade)}$$

Equation F-4b may now be written as

$$\Delta e_{out} = K_1 \log 10 \tag{F-5}$$

The logarithmic slope is

$$LS = \frac{K_1 \text{ volts}}{\text{decade}} \tag{F-6}$$

One decade equals 20 dBV (20 log 10 = 20 dBV), and Equation F-6 may be written as

$$LS = \frac{K_1}{20}\left(\frac{\text{volts}}{\text{dBV}}\right) \tag{F-7}$$

The input to a logarithmic amplifier is often expressed in dBV (dB referenced to 1 volt):

$$e_{in}(\text{dBV}) = 20 \log e_{in} \tag{F-8a}$$

or

$$e_{in} = 10^{\frac{e_{in}(\text{dBV})}{20}} \tag{F-8b}$$

Equation F-1 may now be written as

$$e_{out} = K_1 \log\left(10^{\frac{e_{in}(\text{dBV})}{20}}\right) + K_1 \log K_2 \tag{F-9a}$$

or

$$e_{out} = \frac{K_1 e_{in}(\text{dBV})}{20} + K_1 \log K_2 \tag{F-9b}$$

Thus, from Equations F-9b and F-7 and Figure F-1b, it may be seen that the output increases $K_1/20$ volts for each dBV increase in the input.

Varying K_2 will increase the output by a constant term (Equation F-2) and also will shift the value of e_{in} for $e_{out} = 0$. The value of e_{in} for $e_{out} = 0$ may easily be found by solving Equation F-1 with $e_{out} = 0$.

$$e_{out} = K_1 \log K_2 \, e_{in} = 0 \tag{F-10}$$

or

$$e_{in} = \frac{1}{K_2} \tag{F-11a}$$

If K_2 is given in dBV (20 log K_2), the input for $e_{out} = 0$ is

$$e_{in}(\text{dBV}) = -K_2(\text{dBV}) \tag{F-11b}$$

The incremental linear gain for the logarithmic amplifier is needed when solving for the loop gain and static regulation. This incremental gain may be given as

$$A_{log} = \frac{de_o}{de_{in}} \quad (e_{in} = \text{constant}) \tag{F-12}$$

or

$$A_{log} = K_1 \, (\log \epsilon) \frac{1}{e_{in}} = \frac{0.43 \, K_1}{e_{in}} \tag{F-13}$$

However, substituting Equation F-7 for K_1,

$$A_{log} = \frac{8.6 \, (LS)}{e_{in}} \tag{F-14}$$

Nomenclature

A	Conventional gain term
A_{AGC}	Dynamic AGC gain
$A_{D,Lin}$	Dynamic detector gain for linear detector
$A_{D,SL}$	Dynamic detector gain for square law detector
A_{IF}(dBm)	IF amplifier gain, in dBm
A_{Int}	Integrator frequency dependent gain
A_0(dBm)	Maximum IF amplifier gain, in dBm
A_{PD}	Predetector amplifier gain
A_v	Video amplifier voltage gain
A_Δ	Differencing amplifier gain
A_ϵ	Error amplifier voltage gain
B	Conventional feedback factor
BW	Antenna beamwidth, in degrees
CR	Compression ratio
D	Integrator or low-pass-filter loop update duty cycle
dBm	Decibel referenced to 1 milliwatt
dBV	Decibel referenced to 1 volt
E_{Ref}	AGC reference voltage
e_D	Detector output voltage
e_{DN}	Normalized detector output voltage
$e_{IF,PP}$	Peak-to-peak IF amplifier output voltage
e_N	Normalized video output voltage

$e_{PD,PP}$	Peak-to-peak detector input voltage
e_{rms}	Root mean squared voltage
e_{V}	Video amplifier output voltage
f	Frequency in hertz
$f_{3\mathrm{dBV}}(LG)$	Loop gain 3 dBV frequency response
$f_{3\mathrm{dBV}}(LPF)$	Low-pass-filter 3 dBV frequency response
IMR	Input modulation reduction
I_T	Constant current source in milliamperes
j	$\sqrt{-1}$
K_1, K_2	Logarithmic video amplifier constants
K_{Lin}	Linear detector diode constant
K_{SL}	Square law detector diode constant
LG	Loop gain
LG_{Lin}	Loop gain for linear detector
LG_{SL}	Loop gain for square law detector
LS	Logarithmic amplifier slope, in V/dBV
M_{o}	Output modulation
M_I	Input modulation
M_{IF}	IF amplifier output modulation
N	Number of pulses needed to normalize loop from 10 to 90%
P	Power, in watts
p	Number of pulse returns for a scanning beam radar
$P_{IF,\max}(\mathrm{dBm})$	Maximum IF amplifier signal output power, in dBm, for AGC action
$P_{IF,\min}(\mathrm{dBm})$	Minimum IF amplifier signal output power, in dBm, for AGC action
$P_{\mathrm{in}}(\mathrm{dBm})$	Signal input power, in dBm
$P_{\mathrm{in,max}}(\mathrm{dBm})$	Maximum signal input power, in dBm, for AGC action

$P_{in,min}$(dBm)	Minimum signal input power, in dBm, for AGC action (sometimes referred to as AGC delay).
P_o(dBm)	Signal output power, in dBm
P_{oN}(dBm)	Normalized signal output power, in dBm
P_{PD}(dBm)	Detector input power, in dBm
PRF	Pulse repetition frequency
PRI	Pulse repetition interval
R_c	Collector resistor
S	Laplacian S
T_u	Integrator or low-pass-filter AGC loop update time
T_{pulse}	Integrator or low-pass-filter time constant
X	Slope of variable gain amplifier, dBm/V for IF amplifiers; dBV/V for video amplifiers
Z_F	Operational amplifier feedback impedance
ΔAGC	Change in AGC voltage into IF amplifier
$\Delta AGC'$	Change in AGC voltage at error amplifier output
$\Delta e_{IF,PP}$	Peak-to-peak change in IF amplifier output voltage
Δe_N(dBV)	Change in normalized video output in dBV
$\Delta e_{N,Lin}$(dBV)	Change in normalized video voltage, in dBV, for linear detector
$\Delta e_{N,SL}$(dBV)	Change in normalized video voltage, in dBV, for square law detector
$\Delta e_{PD,PP}$	Peak-to-peak change in detector input voltage
$\Delta e_{\epsilon,PP}$	Peak-to-peak change in error voltage
ΔP_{IF}(dBm)	Change in IF amplifier output power, in dBm
ΔP_{in}(dBm)	Change in input power, in dBm
$\Delta P_{IF,Lin}$(dBm)	Change in IF amplifier output power, in dBm for linear detector
$\Delta P_{IF,SL}$(dBm)	Change in IF amplifier output power, in dBm, for square law detector

ΔP_o(dBm)	Change in output power, in dBm
θ	Scanning beam radar scan width, in degrees
ϕ	Conical scan modulation phase shift, in degrees
$\tau_{r,Lin}$	Loop 10 to 90% rise time for linear detector
$\tau_{r,SL}$	Loop 10 to 90% rise time for square law detector

Bibliography

Fairchild Semiconductor Co. *An Integrated Circuit AGC IF Amplifier.* Mountain View, Calif., FSC, June 1971. Application Note 204.

Locke, A. S. *Guidance.* New York, D. Van Nostrand Co., Inc., 1955, pp. 402-08.

Pogge, R. D. "Designers' Guide to: Basic AGC Amplifier Design," *EDN* (January 20, 1974), pp. 72-76.

Rheinfelder, W. A. "Designing Automatic Gain Control Systems. Part 1. Design Parameters," *EEE* (December 1964), pp. 43-47.

__________. "Designing Automatic Gain Control Systems. Part 2. Circuit Design," *EEE* (January 1965), pp. 53-57.

Victor, W. K., and M. H. Brockman. "The Application of Linear Servo Theory to the Design of AGC Loops," in *Proceedings of the IRE.* (February 1960), pp. 234-38.

Weldon, L. A. "Designing AGC for Transistorized Receivers, Part 1," *Electron Design* (September 13, 1962), p. 64.

__________. "Designing AGC for Transistorized Receivers, Part 2," *Electron Design* (October 11, 1974), p. 76.

NWC 1652 (8/77) 280 SD

www.ingramcontent.com/pod-product-compliance
Lightning Source LLC
LaVergne TN
LVHW091641100826
845152LV00006B/125/J
9781427615756